Grundlagen der
exakten Naturwissenschaften
Band 5

Grundlagen der exakten Naturwissenschaften

Wissenschaftstheoretische, logische und philosophische Untersuchungen zur Physik und den angrenzenden Gebieten

Herausgegeben von Peter Mittelstaedt, Universität Köln

Band 1:
Grundlagen der Quantentheorie
Herausgegeben von P. Mittelstaedt und J. Pfarr

Band 2:
I. Strohmeyer, Transzendentalphilosophische und physikalische Raum-Zeit-Lehre

Band 3:
P. Mittelstaedt, der Zeitbegriff in der Physik

Band 4:
Protophysik und Relativitätstheorie
Herausgegeben von J. Pfarr

Band 5:
Interpretations and Foundations of Quantum Theory
Edited by H. Neumann

Interpretations and Foundations of Quantum Theory

Proceedings of a Conference hold in
Marburg 28–30 May 1979

edited
by Holger Neumann,
Universität Marburg

Bibliographisches Institut Mannheim/Wien/Zürich
B.I.-Wissenschaftsverlag

CIP-Kurztitelaufnahme der Deutschen Bibliothek

Interpretations and foundations of quantum theory: proceedings of a conference hold in Marburg, 28 - 30 May 1979 / ed. by Holger Neumann. - Mannheim; Wien; Zürich: Bibliographisches Institut, 1981.
(Grundlagen der exakten Naturwissenschaften; Bd. 5)
ISBN 3-411-01601-9
NE: Neumann, Holger [Hrsg.]; GT

Satz: Copo-Satz GmbH, Seeheim-Jugenheim 2
Druck und Bindearbeit: Hain-Druck GmbH, Meisenheim/Glan
Printed in Germany
ISBN 3-411-01601-9

PREFACE

A colloquium "Interpretations and Foundations of Quantum Theory" was held at Marburg, Germany from May 28–30, 1979. The aim of this colloquium was to stimulate the discussion of the formal and semantical relations between different approaches to the foundation of quantum theory. There are several well established approaches to the foundations of quantum theory differing with respect to the physical notions considered as basic for the theory, the epistemology and the mathematical methods applied. Though there is an amount of literature on these approaches, few is known concerning their connections and differences.

This volume contains articles based on lectures of the colloquium. It consists of contributions to four different approaches to the foundations of quantum theory and includes two contributions in which relations between these approaches are discussed. However, the discussion of the relations can only be considered as opened by these contributions.

I would like to thank the Deutsche Forschungsgemeinschaft who completely sponsored the colloquium.

Marburg, April 1980

Holger Neumann

CONTENTS

Empirical Logic and Tensor Products

by

D. J. Foulis and C. H. Randall

1. Introduction

In our work we are developing a formalism called *empirical logic* to support a generalization of conventional statistics; the resulting generalization is called *operational statistics*. We are not attempting to develop or advocate any particular physical theory; rather we are formulating a precision "language" in which such theories can be expressed, compared, evaluated, and related to laboratory experiments. We believe that only in such a language can the connections between real physical procedures (operations) and physical theories be made explicit and perspicuous.

Although our formalism allows us to distinguish between that which is operationally accessible and that which is not, whenever we choose to do so, we are not operationalists, logical positivists, or radical empiricists. In fact our formalism, as far as possible, is meant to be independent of any particular epistemological point of view. For instance, it neither requires nor discourages the realist view of the world. In [7], we have compared and contrasted our approach with alternative approaches to the foundations of empirical science.

In our research program we always hold before us two paradigms or benchmarks: The Kolmogorov foundation for probability and statistics as set forth in [5] and the Dirac-von Neumann foundation for orthodox quantum mechanics as set forth in [1] and [9].

2. Basic Definitions

When we speak of a *physical operation*, we have in mind a set of instructions for executing a well-defined, physically realizable, reproducible procedure. These instructions must include the specification of a definite set of symbols corresponding to the possible outcomes of the procedure. Each execution of the procedure must result in one and only one symbol in this outcome set. The symbols used to represent outcomes must be so constituted that the operation (or operations) capable of yielding these outcomes can be determined from the makeup of the symbols themselves. Thus, a physical operation can and will be represented by its own outcome set.

Outcome sets corresponding to different physical operations can overlap. The idea is that outcomes of different physical operations are to be identified if they are physically equivalent in some sense. The decision as to just which outcomes are to be regarded as being physically equivalent could be based on a variety of considerations ranging from an attractive whim to an elaborate scientific theory – often such a decision is inspired by a particular world picture or model. In many cases, the instructions for two distinct physical operations may all but dictate the identification of some outcomes; see, for instance, Example III on page 1674 of [2]. The equivalence of motor vehicle speeds as measured by a speedometer on the one hand and radar on the other provides a familiar example of outcome identification, one that is, in fact, upheld by courts of law.

When we speak of a *manual* of physical operations, we have in mind a collection or catalogue of experiments, usually cohesive and complete in some sense, as, for example, a laboratory manual. Since we have agreed to represent a physical operation by its outcome set, a manual of physical operations will be represented by a collection of nonempty sets which, in general, will overlap for the reasons explained above. In order to avoid "grammatical chaos" in our language, certain mild normative conditions are required for manuals.

2.1 Definition. *A quasimanual* A is a nonempty set of nonempty sets. By an A*-operation* we mean any set $E \in \mathsf{A}$. Bv an A*-outcome* we mean any element $e \in E$, where E is any A-operation. By an A*-event* we mean any subset $A \subseteq E$, where E is any A-operation. We denote the set of all A-events by $\mathsf{E}(\mathsf{A})$.

Henceforth, we understand A to be a quasimanual.

2.2 Definition. If A and B are A-events for which $A \cap B = \phi$ and such that $A \cup B$ is again an A-event, then we say that A is *orthogonal* to B and we write $A \perp B$. If, in addition to $A \perp B$, the event $A \cup B$ is an A-operation, then we say that A and B are *operational complements* and we write A oc B.

The terminology in the following definition is motivated by [10, p. 16].

2.3 Definition. If A and B are A-events that share a common operational complement C in $\mathsf{E}(\mathsf{A})$, then we say that A and B are *operationally perspective* with *axis* C and we write A op B. If there exists a finite sequence of A-events $D_0, D_1, \ldots, D_n$ such that $A = D_0, B = D_n$, and D_{i-1} op D_i for $i = 1, 2, \ldots, n$, then we say that A is *weakly perspective* to B and we write A wp B. We denote by $p(A)$ the set of all A-events B such that B wp A.

3. Occurrence, Nonoccurrence, Confirmation, and Refutation

Let A be an A-event, so that, by Definition 2.1, there exists at least one A-operation E containing A. Such an E is called a *test operation* for A. If a test operation for A is executed, then A is said to *occur* or to *nonoccur* according to whether the outcome secured belongs or does not belong to

A. Now, suppose that B is a second $\mathfrak{A}$-event with A oc B and let $E = A \cup B$. If E is executed and A occurs, then obviously B nonoccurs. However, if we are told only that A occurs, we cannot conclude that B nonoccurs. Indeed, there may be test operations for A that are not test operations for B. Although it may seem like hairsplitting to insist that an occurrence of A need not imply a nonoccurrence of the complementary event B, we believe that it is precisely this insistence that is crucial to a proper understanding of operational logic in general and quantum physics in particular.

The erroneous notion that an occurrence of A entails a nonoccurrence of the complementary event B is probably rooted in the realist view of the world, which would cause us to regard A and B as contradictory properties of some real physical object or system – the presence of one property would automatically imply the absence of the other. Although our formalism does not require a realist view of the world, it should not discourage such a view either. Thus, we propose to recognize that,if A occurs, then there is some weak sense in which B fails to occur. Therefore, if A oc B and A occurs, then we shall say that B is *refuted*; similarly, if A nonoccurs, then we shall say that B is *confirmed.* Moreover, as a natural extension of these ideas, if A oc B and A is confirmed (respectively, refuted), let us agree to say that B is *refuted* (respectively, *confirmed*).

As an immediate consequence of the agreements made above, it is clear that the occurrence, or confirmation, of an event A entails a confirmation of all the events in the equivalence class $p(A)$ and, simultaneously, a refutation of all the events in the equivalence class $p(B)$ for any B oc A. Likewise, if A nonoccurs, or is refuted, then all events in $p(A)$ are refuted and all events in $p(B)$ are confirmed. These considerations suggest the following definition.

3.1 Definition. For $A \in \mathcal{E}(\mathfrak{A})$, the *negation* of $p(A)$, in symbols $p(A)'$, is defined by $p(A)' = p(B)$, where B is any $\mathfrak{A}$-event for which B oc A.

If A is an $\mathfrak{A}$-event, then any test operation for A will be referred to as a *test operation* for the *operational proposition* $p(A)$. If such a test operation is executed, we say that $p(A)$ is *confirmed* or *refuted* according to whether A occurs or nonoccurs. Note that $p(A)$ is confirmed if and only if its negation $p(A)'$ is refuted and vice versa.

There are many other modes of confirmation and refutatation, suggested by the canons of classical reasoning, that clamor for possible adoption. On close examination they are all found to require that the quasimanual $\mathfrak{A}$ satisfy various regularity conditions. We shall consider these matters at length in another paper. Here we have elected to give a less general and somewhat simpler exposition, and thus will consider only manuals – quasimanuals that satisfy the following normative condition.

3.2 Definition. $\mathfrak{A}$ is a *manual* if and only if, for $A, B \in \mathcal{E}(\mathfrak{A})$, $B \in p(A)'$ implies that B oc A.

If $\mathfrak{A}$ is a manual, $A_1 \in p(A)$, $B_1 \in p(B)$, and $A_1 \perp B_1$, it follows that $A \perp B$; hence, we can make the following definition.

3.3 Definition. Let $\mathfrak{A}$ be a manual and let $A, B \in \mathcal{E}(\mathfrak{A})$. If $A \perp B$, we say that $p(A)$ is *orthogonal* to $p(B)$ and we write $p(A) \perp p(B)$. If $p(A) \perp p(B)$, we define the *orthogonal sum* $p(A) \oplus p(B) = p(A \cup B)$.

4. Operational Logics

4.1 Definition. If $\mathfrak{A}$ is a manual, we define the *operational logic* of $\mathfrak{A}$ to be the system $(\Pi(\mathfrak{A}), \perp, \oplus, ', 0, 1)$ where $\Pi(\mathfrak{A}) = \{p(A) \mid A \in \mathcal{E}(\mathfrak{A})\}$, $\perp$, $\oplus$, and $'$ are as previously defined, $0 = p(\phi)$, and $1 = 0'$.

Such logics are easily characterized in terms of the following:

4.2 Definition. An *operational logic* $(L, \perp, \oplus, ', 0, 1)$ is a system consisting of a set L, a binary relation $\perp$ on L, a partially defined binary operation $\oplus$ on L, a unary operation $p \mapsto p'$ on L, and two special elements 0 and 1 of L satisfying the following conditions for all elements $p, q, r \in L$:

(i) (orthogonal sums)	$p \oplus q$ is defined $\Leftrightarrow p \perp q$.
(ii) (commutativity)	$p \perp q \Rightarrow q \perp p$ and $p \oplus q = q \oplus p$.
(iii) (associativity)	$p \perp q, (p \oplus q) \perp r \Rightarrow q \perp r$, $p \perp (q \oplus r)$, and $p \oplus (q \oplus r) = (p \oplus q) \oplus r$.
(iv) (orthocomplements)	For every $p \in L$, there exists a unique $q \in L$, namely $q = p'$, such that $p \perp q$ and $p \oplus q = 1$.
(v) (maximality of 1)	$p \perp 1 \Rightarrow p = 0$.
(vi) (minimality of 0)	$p \perp 0$ and $p \oplus 0 = p$.
(vii) (consistency)	$p \perp p \Rightarrow p = 0$.

By an *operational sublogic* of L, we mean a subset Q of L, containing 0 and 1, closed under $'$, and closed under $\oplus$ in the sense that, if $p, q \in Q$ with $p \perp q$, then $p \oplus q \in Q$.

Evidently, the operational logic of a manual satisfies the conditions of Definition 4.2. Conversely, it can be shown that every operational logic L is isomorphic to the operational logic $\Pi(\mathfrak{A})$ of a suitable manual $\mathfrak{A}$. If one wishes to study confirmation and refutation, as opposed to the more stringent requirements entailed by occurrence and nonoccurrence, then one naturally identifies operationally perspective events and analyzes only the logic $\Pi(\mathfrak{A})$ at the cost of discarding information that could prove to be crucial.

4.3 Definition. An operational logic L is said to be *orthocoherent* if $p, q, r \in L$ with $p \perp q$, $p \perp r$, and $q \perp r$ implies that $(p \oplus q) \perp r$.

Any orthomodular partially ordered set (OMP) is an orthocoherent operational logic if the orthogonal sum is understood to be the orthogonal join. Conversely, given any orthocoherent operational logic L, there exists a unique partial order relation $\leqslant$ on L such that $(L, \leqslant, \perp, ', 0, 1)$ is an OMP for which the orthogonal join is the orthogonal sum. In this sense, an orthocoherent operational logic is the same thing as an OMP.

It can be shown that every operational logic L is a union of orthocoherent sublogics each of which, as noted above, carries its own unique partial order $\leqslant$. Moreover, these partial orders agree on the intersections of such sublogics. This suggests defining a global partial order $\leqslant$ on L. The partial order relation in question is, in fact, provided by the following definition.

4.4 Definition. If p and q are elements of an operational logic L, we define $p \leqslant q$ to mean that there exists an element $r \in L$ with $p \perp r$ and $p \oplus r = q$.

Although $p \oplus q$ is the supremum of p and q in any orthocoherent sublogic containing both p and q, it need not be effective as a supremum of p and q in L. In fact, the condition that, for $p, q \in L$ with $p \perp q$, $p \oplus q$ is always the supremum of p and q with respect to the partial order relation of Definition 4.4 is equivalent to the condition that L be orthocoherent, and thus an OMP.

A manual $\mathcal{A}$ is called *orthocoherent* if its logic $\Pi(\mathcal{A})$ is orthocoherent. In our earlier papers [2], [6], [3], we routinely imposed a condition called *coherence* on our manuals. Coherence, in fact, implies orthocoherence, and is equivalent to it in manuals possessing only finite operations or only a finite number of operations.

5. States

Let X denote the set of all outcomes of the quasimanual $\mathcal{A}$.

5.1 Definition. An $\mathcal{A}$-*weight* is a function $\omega : X \longrightarrow [0, 1]$ such that, for every $E \in \mathcal{A}$ $\sum_{x \in E} \omega(x) = 1$, the sum being understood in the unordered sense. The set of all $\mathcal{A}$-weights is denoted by $\Omega = \Omega(\mathcal{A})$. For $\omega \in \Omega$ and $A \in \mathcal{E}(\mathcal{A})$, define $\omega(A) = \sum_{x \in A} \omega(x)$. The resulting map $\omega : \mathcal{E}(\mathcal{A}) \longrightarrow [0, 1]$ is called a *regular* $\mathcal{A}$-*state*.

Notice that, if $\omega \in \Omega$ and if A and B are $\mathcal{A}$-events with $A \perp B$, then $\omega(A \cup B) = \omega(A) + \omega(B)$.

5.2 Definition. A function $\alpha : \mathcal{E}(\mathcal{A}) \longrightarrow [0, 1]$ such that $\alpha(E) = 1$ for all $E \in \mathcal{A}$ and such that $\alpha(A \cup B) = \alpha(A) + \alpha(B)$ whenever A and B are $\mathcal{A}$-events with $A \perp B$ is called an $\mathcal{A}$-*state*.

Notice that, if α is an $\mathcal{A}$-state, then α assigns the same value to all $\mathcal{A}$-events in an equivalence class $p(A)$.

5.3 Definition. Let α be an $\mathcal{A}$-state and let $A \in \mathcal{E}(\mathcal{A})$. Then we define $\alpha(p(A)) = \alpha(A)$.

5.4 Definition. By a *state* α on an operational logic L, we mean a function $\alpha : L \longrightarrow [0, 1]$ such that $\alpha(1) = 1$ and such that $\alpha(p \oplus q) = \alpha(p) + \alpha(q)$ whenever $p \perp q$ in L.

Notice that, if $\mathcal{A}$ is a manual, then the $\mathcal{A}$-states can be identified with the states on the logic $\Pi(\mathcal{A})$.

A regular $\mathcal{A}$-state ω is to be regarded as a logically possible complete stochastic model for the empirical situation described by the quasimanual $\mathcal{A}$. This is to be understood in the sense that, for every $\mathcal{A}$-event A, $\omega(A)$ is interpreted as the "long run relative frequency" with which the event A occurs as a consequence of executions of test operations. Therefore, $\omega(p(A))$ is the frequency with which $p(A)$ is confirmed when tested. Although this provides an interpretation for the regular states on the logic $\Pi(\mathcal{A})$, there is no way, in general, to tell just which states on an operational logic are regular unless the underlying manual $\mathcal{A}$ is known. This is indicative of the manner in which vital information may be lost in the passage from a manual $\mathcal{A}$ to its logic $\Pi(\mathcal{A})$.

In view of the interpretation of regular states as complete stochastic models for $\mathcal{A}$, virtually any version of the controversial projection postulate would require that $\Omega = \Omega(\mathcal{A})$ be "unital" in the following sense.

5.5 Definition. A set Δ of $\mathcal{A}$-states is called unital provided that, for every nonempty $\mathcal{A}$-event A, there exists an $\alpha \in \Delta$ such that $\alpha(A) = 1$.

6. Interpretation Morphisms and Premanuals

6.1 Definition. Let $\mathcal{A}$ and $\mathcal{B}$ be quasimanuals and let X denote the set of all $\mathcal{A}$-outcomes. By an *interpretation morphism* $\mathcal{A} \xrightarrow{\Phi} \mathcal{B}$, we mean a map $\Phi : X \longrightarrow \mathcal{E}(\mathcal{B})$ with the following two properties:

(i) For $E \in \mathcal{A}$, $\bigcup_{x \in E} \Phi(x) \in \mathcal{B}$.

(ii) Any two distinct $\mathcal{A}$-outcomes belonging to the same $\mathcal{A}$-operation are mapped by Φ to disjoint $\mathcal{B}$-events.

For such a Φ, if $A \subseteq X$, define $\Phi(A) = \bigcup_{x \in A} \Phi(x)$. We call Φ *positive* in case $\Phi(x) \neq \phi$ for all $x \in X$.

If $\mathcal{A} \xrightarrow{\Phi} \mathcal{B}$ is an interpretation morphism and $A \perp B$ in $\mathcal{E}(\mathcal{A})$, then, clearly, $\Phi(A) \perp \Phi(B)$ in $\mathcal{E}(\mathcal{B})$. Moreover, if A oc B, then $\Phi(A)$ oc $\Phi(B)$, and, consequently, if A wp B, then $\Phi(A)$ wp $\Phi(B)$. Also notice that $\omega \in \Omega(\mathcal{B})$ implies $\omega \circ \Phi \in \Omega(\mathcal{A})$.

6.2 Definition. Let $\mathcal{A} \xrightarrow{\Phi} \mathcal{B}$ be an interpretation morphism. We define the *conjugate map* $\Phi^\dagger : \Omega(\mathcal{B}) \longrightarrow \Omega(\mathcal{A})$ by the equation $\Phi^\dagger(\omega) = \omega \circ \Phi$ for all $\omega \in \Omega(\mathcal{B})$.

We use the term "interpretation" because we have in mind the situation in which $\mathcal{A}$ is a phenomenological or laboratory quasimanual explained or interpreted by the theoretical or model quasimanual $\mathcal{B}$. Thus, an $\mathcal{A}$-outcome x and its corresponding $\mathcal{B}$-event $\Phi(x)$ are understood to represent the "same effect." In practice, of course, the model quasimanual $\mathcal{B}$ will have considerably more mathematical structure than the laboratory quasimanual $\mathcal{A}$ – for instance, $\mathcal{B}$ may be the coherent manual of all countable Borel measurable partitions of a classical phase space. In particular, notice that the conjugate map $\Phi^\dagger$ interprets a theoretical regular state ω as a laboratory frequency model $\Phi^\dagger(\omega)$. Naturally, $\Phi^\dagger$ need not be one-to-one or onto. There may be many theoretical states that predict the same frequencies for the laboratory events. Also, there may be regular $\mathcal{A}$-states that are not consistent with the interpretation Φ.

6.3 Definition. A quasimanual $\mathcal{A}$ is called a *premanual* if there exists a manual $\mathcal{B}$ and a positive interpretation morphism $\mathcal{A} \xrightarrow{\Phi} \mathcal{B}$.

It can be shown that a quasimanual $\mathcal{A}$ is a premanual if and only if it is a subset of a manual $\mathcal{B}$. Consequently, we can make the following definition:

6.4 Definition. If $\mathcal{A}$ is a premanual, then $\langle\mathcal{A}\rangle$ is defined to be the set-theoretic intersection of all the manuals that contain $\mathcal{A}$.

6.5 Theorem. Let $\mathcal{A}$ be a premanual. Then:

(i) $\langle\mathcal{A}\rangle$ is a manual.
(ii) $\mathcal{A} \subseteq \langle\mathcal{A}\rangle$ and, if $\mathcal{B}$ is a manual, $\mathcal{A} \subseteq \mathcal{B}$ implies that $\langle\mathcal{A}\rangle \subseteq \mathcal{B}$.
(iii) The $\langle\mathcal{A}\rangle$-outcomes are the same as the $\mathcal{A}$-outcomes.
(iv) $\Omega(\langle\mathcal{A}\rangle) = \Omega(\mathcal{A})$.

6.6 Theorem. *If* $\mathcal{A}$ *is a quasimanual such that* $\Omega(\mathcal{A})$ *is unital, then* $\mathcal{A}$ *is a premanual.*

7. Examples

We now give four examples to illustrate the ideas introduced above.

7.1 Example – The "Hook". The simplest example of a premanual that is not a manual is the *hook* $\mathcal{K} = \{\{a, b\}, \{b, c\}, \{c, d\}\}$. Notice that $\langle\mathcal{K}\rangle = \{\{a, b\}, \{b, c\}, \{c, d\}, \{d, a\}\}$, and $\Pi(\langle\mathcal{K}\rangle)$ is the Boolean algebra with four elements. Here the state space $\Omega(\mathcal{K}) = \Omega(\langle\mathcal{K}\rangle)$ is one-dimensional.

7.2 Example – The "Wright Triangle." The simplest example of a manual that is not coherent is the Wright triangle $\mathcal{W} = \{\{a, b, c\}, \{c, d, e\}, \{e, f, a\}\}$. The logic $\Pi(\mathcal{W})$ consists of three eight-element Boolean algebras "pasted together" in a loop. It is not an orthomodular partially ordered set. The state space $\Omega(\mathcal{W})$ is three dimensional and contains a unital subset of *dispersion free* or *deterministic* states (that is, states that assume only the values 0 and 1). Notice that the state that assigns values 1/2 to outcomes a, c, and e is *pure* (that is, it cannot be written as a convex combination of other states), but it is not dispersion free.

7.3 Example – "Borel Manuals." Orthodox probability theory and classical statistical mechanics both depend on the notion of a *Borel space* $(S, \mathcal{M})$ consisting of a nonempty set S and a σ-field $\mathcal{M}$ of subsets of S. For such a Borel space $(S, \mathcal{M})$, let us define the associated *Borel manual* $\mathcal{B} = \mathcal{B}(S, \mathcal{M})$ to be the set of all countable $\mathcal{M}$-measurable partitions of S. We regard $\mathcal{B}$-operations to be "coarsenings" of a (possibly) idealized physical operation with outcome set S. The manual $\mathcal{B}$ is coherent and its logic $\Pi(\mathcal{B})$ is canonically isomorphic to the σ-complete Boolean algebra $\mathcal{M}$. The regular states in $\Omega(\mathcal{B})$ are precisely the countably additive probability measures for the Borel space $(S, \mathcal{M})$. Incidently, the $\mathcal{B}$-states that are not necessarily regular are just the finitely additive probability measures for $(S, \mathcal{M})$.

7.4 Example – "Frame Manuals." Orthodox quantum mechanics depends on the notion of a Hilbert space $\mathcal{H}$. For such a Hilbert space $\mathcal{H}$, let us define the associated *frame manual* $\mathcal{F}(\mathcal{H})$ to be the set of all orthonormal bases (frames) for $\mathcal{H}$. The manual $\mathcal{F}(\mathcal{H})$ is coherent and its logic $\Pi(\mathcal{F}(\mathcal{H}))$ is canonically isomorphic to the complete orthomodular lattice of projection operators on $\mathcal{H}$. The regular states in $\Omega(\mathcal{F}(\mathcal{H}))$ are, according to Gleason's theorem, in one-to-one correspondence with the von Neumann density operators on $\mathcal{H}$.

An important example of a Borel interpretation of a quasimanual $\mathcal{A}$ for which $\Omega_{df}(\mathcal{A})$, the set of all regular dispersion free $\mathcal{A}$-states, is nonempty can be obtained as follows: Let $S = \Omega_{df}(\mathcal{A})$ and, for any $\mathcal{A}$-outcome x, let $x^1 = \{\omega \in \Omega_{df}(\mathcal{A}) | \omega(x) = 1\}$. Let $\mathcal{M}$ be the σ-field of subsets of S generated by all sets of the form x^1 as x runs through all $\mathcal{A}$-outcomes. If $x^1 \neq \phi$, define $\Phi(x) = \{x^1\}$; otherwise, define $\Phi(x) = \phi$. Then $\mathcal{A} \xrightarrow{\Phi} \mathcal{B}(S, \mathcal{M})$ is an interpretation morphism. Notice that Φ is positive if and only if $\Omega_{df}(\mathcal{A})$ is unital. For the case of coherent manuals, additional facts about morphisms can be found in [4], [7], and [11].

8. Cartesian and Operational Products

In many cases the elements of a quasimanual represent the physical operations that a single observer many execute. In other instances, the operations in a quasimanual represent the observations or measurements that are permitted on a given type of physical system. The problem of representing the combined activities of two observers, or equivalently the problem of representing the union of two physical systems, requires the introduction of a suitable "product" of quasimanuals. In this and the following section, we shall consider three such constructions: *Cartesian, operational*, and *tensor* products. In a companion paper [8], it will be shown that these are, respectively, the products to use when one wishes to represent *bilateral, unilateral*, or *no* influence (or interference) between the two quasimanuals involved.

In what follows, let $\mathcal{A}$ and $\mathcal{B}$ denote quasimanuals with outcome sets X and Y, respectively. If $x \in X$ and $y \in Y$, we write the ordered pair (x, y) simply as xy. Similarly, if $A \subseteq X$ and $B \subseteq Y$, we write the cartesian product $A \times B$ simply as AB. Of course, xB will mean $\{x\}B$.

8.1 Definition. We define the *cartesian product* of $\mathcal{A}$ and $\mathcal{B}$ to be the quasimanual $\mathcal{A} \times \mathcal{B} = \{EF \mid E \in \mathcal{A}, F \in \mathcal{B}\}$.

An execution of an operation EF in $\mathcal{A} \times \mathcal{B}$ would presumably take place as follows: Operation E in $\mathcal{A}$ is executed to secure (say) outcome $x \in E$, operation F in $\mathcal{B}$ is executed to secure (say) outcome $y \in F$, and the outcome of this execution of the compound operation EF in $\mathcal{A} \times \mathcal{B}$ is duly recorded as xy.

The cartesian product of two manuals is again a manual; likewise for premanuals, coherent manuals, or orthocoherent manuals. If $\mathcal{A}$ and $\mathcal{B}$ are manuals, in general it is not possible to determine the structure of the logic $\Pi(\mathcal{A} \times \mathcal{B})$ from the structures of the logics $\Pi(\mathcal{A})$ and $\Pi(\mathcal{B})$. Again, vital information is lost in the passage from manuals to their logics.

8.2 Definition. A set of the form $\bigcup_{x \in E} xF_x$, where E is an $\mathcal{A}$-operation and F_x is a $\mathcal{B}$-operation for every $x \in E$, is called a *two-stage* operation. The quasimanual consisting of all such two-stage operations is called the *(forward) operational product of* $\mathcal{A}$ *and* $\mathcal{B}$ and is denoted by $\overrightarrow{\mathcal{A}\mathcal{B}}$.

An execution of a two-stage operation $\bigcup_{x \in E} xF_x$ in $\overrightarrow{\mathcal{A}\mathcal{B}}$ would presumably take place as follows: Operation E in $\mathcal{A}$ is executed to secure (say) outcome $x \in E$, and then operation F_x in $\mathcal{B}$ is executed to secure (say) outcome $y \in F_x$; the outcome of this execution of the two-stage operation being duly recorded as xy.

8.3 Definition. If $S \subseteq YX$, we define $S^{-1} = \{xy \mid yx \in S\}$. The *(backward) operational product of* $\mathcal{A}$ *and* $\mathcal{B}$ is defined to be the quasimanual $\overleftarrow{\mathcal{A}\mathcal{B}} = \{S^{-1} \mid S \in \overrightarrow{\mathcal{B}\mathcal{A}}\}$.

Thus, to execute an operation S^{-1} in $\overleftarrow{\mathcal{A}\mathcal{B}}$, we would execute the two-stage operation S in $\overrightarrow{\mathcal{B}\mathcal{A}}$ to secure (say) the outcome yx, but we would record this $\overleftarrow{\mathcal{A}\mathcal{B}}$-outcome as xy.

The (forward or backward) operational product of two manuals is again a manual; likewise for premanuals, coherent manuals, or orthocoherent manuals. If $\mathcal{A}$ and $\mathcal{B}$ are manuals, the structure of the logic $\Pi(\overrightarrow{\mathcal{A}\mathcal{B}})$ depends on the detailed structure of the manual $\mathcal{A}$, but only on the structure of the logic $\Pi(\mathcal{B})$.

Notice that all three quasimanuals $\mathcal{A} \times \mathcal{B}$, $\overrightarrow{\mathcal{A}\mathcal{B}}$, and $\overleftarrow{\mathcal{A}\mathcal{B}}$ have the same outcome set, namely XY. Evidently, the cartesian product is contained in both the forward and the backward operational product; in fact, $\mathcal{A} \times \mathcal{B} = \overrightarrow{\mathcal{A}\mathcal{B}} \cap \overleftarrow{\mathcal{A}\mathcal{B}}$.

9. Tensor Products

In this section, let $\mathcal{A}$ and $\mathcal{B}$ be manuals with outcome sets X and Y, respectively.

9.1 Definition. $\overleftrightarrow{\mathcal{A}\mathcal{B}} = \overrightarrow{\mathcal{A}\mathcal{B}} \cup \overleftarrow{\mathcal{A}\mathcal{B}}$.

The quasimanual $\mathcal{A}\mathcal{B}$ turns out to be a manual only in the most trivial cases; however, it often is a premanual. As a consequence of Theorem 6.6, we have the following result.

9.2 Lemma. If $\Omega(\mathcal{A})$ and $\Omega(\mathcal{B})$ are unital, then $\overleftrightarrow{\mathcal{AB}}$ is a premanual.

9.3 Definition. The manuals $\mathcal{A}$ and $\mathcal{B}$ are said to *have a tensor product* if $\overleftrightarrow{\mathcal{AB}}$ is a premanual, in which case, the *tensor product of* $\mathcal{A}$ *and* $\mathcal{B}$ is defined to be the manual $\mathcal{A} \otimes \mathcal{B} = \langle \overleftrightarrow{\mathcal{AB}} \rangle$

By lemma 9.2, if $\mathcal{A}$ and $\mathcal{B}$ have unital sets of regular states, then they have a tensor product $\mathcal{A} \otimes \mathcal{B}$.

9.4 Definition. Let $\mathcal{C}$ be a quasimanual and let a, b be $\mathcal{C}$-outcomes. We define $a \perp b$ to mean that $\{a\} \perp \{b\}$. We say that $\Omega(\mathcal{C})$ is *pointwise strong* provided that, if $a \perp b$ fails to be true, there exists $\omega \in \Omega(\mathcal{C})$ such that $\omega(a) = 1$ and $\omega(b) \neq 0$.

Since no outcome is orthogonal to itself, pointwise strong implies unital. We have the following result.

9.5 Lemma. Let $\Omega(\mathcal{A})$ and $\Omega(\mathcal{B})$ be pointwise strong. Let a and b be $\mathcal{A}$-outcomes and let c and d be $\mathcal{B}$-outcomes. Then $ac \perp bd$ in $\overleftrightarrow{\mathcal{AB}}$ holds if and only if $ac \perp bd$ in $\mathcal{A} \otimes \mathcal{B}$ holds.

By Theorem 6.5, $\overleftrightarrow{\mathcal{AB}}$ and $\mathcal{A} \otimes \mathcal{B}$ (if it exists) have the same set of outcomes, namely XY. Obviously, $ac \perp bd$ in $\overleftrightarrow{\mathcal{AB}}$ holds if and only if either $a \perp b$ holds in $\mathcal{A}$ or $c \perp d$ holds in $\mathcal{B}$ (or both). Also, if $ac \perp bd$ holds in $\overleftrightarrow{\mathcal{AB}}$, then it automatically holds in $\mathcal{A} \otimes \mathcal{B}$. The content of Lemma 9.5 is that, in the presence of pointwise strong sets of regular states, the converse is true.

The structure of $\mathcal{A} \otimes \mathcal{B}$, and of $\Pi(\mathcal{A} \otimes \mathcal{B})$, can be exceedingly complex – for instance, there are simple examples in which $\mathcal{A}$ and $\mathcal{B}$ contain fewer than five operations while $\mathcal{A} \otimes \mathcal{B}$ contains well over a thousand. Nevertheless, because of the following lemma (which is a direct consequence of Theorem 6.5), it is relatively easy to deal with the regular states on $\mathcal{A} \otimes \mathcal{B}$.

9.6 Lemma. Suppose that the manuals $\mathcal{A}$ and $\mathcal{B}$ have a tensor product. Then

$$\Omega(\mathcal{A} \times \mathcal{B}) \supseteq \Omega(\overrightarrow{\mathcal{AB}}) \cup \Omega(\overleftarrow{\mathcal{AB}}) \supseteq \Omega(\overrightarrow{\mathcal{AB}}) \cap \Omega(\overleftarrow{\mathcal{AB}}) = \Omega(\overleftrightarrow{\mathcal{AB}}) = \Omega(\mathcal{A} \otimes \mathcal{B}) \supseteq \Omega(\mathcal{A})\, \Omega(\mathcal{B}).$$

In the preceding lemma, $\Omega(\mathcal{A})\, \Omega(\mathcal{B})$ is understood to consist of all of the *regular product states*, that is, all states of the form $\alpha\beta$ with $\alpha \in \Omega(\mathcal{A})$ and $\beta \in \Omega(\mathcal{B})$, it being understood that, for $x \in X$ and $y \in Y$, $\alpha\beta(xy) = \alpha(x)\beta(y)$. Of course, such product states exhibit the statistical independence of operations in $\mathcal{A}$ from the operations in $\mathcal{B}$.

As is shown by the following example, *the category of orthocoherent manuals (and therefore, the category of orthomodular partially ordered sets) is too small to support tensor products.*

9.7 Example – "The Tensor Product of Pentagons." Consider the *pentagon* manual $\mathcal{P} = \{a,x,b\}, \{b,y,c\}, \{c,z,d\}, \{d,u,e\}, \{e,v,a\}$. Notice that

$\mathcal{P}$ has a pointwise strong set of dispersion free regular states, it is coherent, and its logic $\Pi(\mathcal{P})$ is, in fact, an orthomodular lattice. The operations in this manual $\mathcal{P}$ can even be interpreted in terms of drawing balls from an urn [12, page 271]. Since $\mathcal{P}$ has a unital set of regular (dispersion free) states, then $\mathcal{P} \otimes \mathcal{P}$ exists by Lemma 9.2. Now let $D = \{aa, bc, ce, db, ed\}$. Notice that the outcomes in D are pairwise orthogonal in $\overleftrightarrow{\mathcal{P}\mathcal{P}}$, hence also in $\mathcal{P} \otimes \mathcal{P}$. (As a matter of fact, using Lemma 9.5, it can be shown that D is a maximal set of pairwise orthogonal outcomes in $\mathcal{P} \otimes \mathcal{P}$.) Since D is a finite set, if $\mathcal{P} \otimes \mathcal{P}$ were orthocoherent, then D would be an event (in fact, it would be an operation) for $\mathcal{P} \otimes \mathcal{P}$. Let α be the regular state for $\mathcal{P}$ that assigns the value 1/2 to $a, b, c, d,$ and e and the value 0 to $x, y, z, u,$ and v. (It can be shown that α is a pure state for $\mathcal{P}$.) By Lemma 9.6, the product state $\alpha\alpha$ belongs to $\Omega(\mathcal{P} \otimes \mathcal{P})$. Thus, if D were a $\mathcal{P} \otimes \mathcal{P}$-event, we would have $\alpha\alpha(D) = 5/4$, a contradiction. Thus, we must conclude that D is not a $\mathcal{P} \otimes \mathcal{P}$-event, $\mathcal{P} \otimes \mathcal{P}$ is not orthocoherent, and the logic $\Pi(\mathcal{P} \otimes \mathcal{P})$ is not an orthomodular partially ordered set.

Essentially the same argument, as in the example above, goes to show that there is no hope of obtaining a general "tensor product" $P \otimes Q$ of orthomodular posets P and Q with the following desirable properties:

(i) $P \otimes Q$ is an *OMP*.

(ii) $\otimes$ is a map from $P \times Q$ to $P \otimes Q$ such that, $p_1 \otimes q_1 \perp p_2 \otimes q_2$ if either $p_1 \perp p_2$ or $q_1 \perp q_2$.

(iii) If α and β are states for P and Q, respectively, then there exists a state γ on $P \otimes Q$ such that $\gamma(p \otimes q) = \alpha(p)\, \beta(q)$ for all $p \in P$ and all $q \in Q$.

In fact, such a $P \otimes Q$ cannot exist for $P = Q = \Pi(\mathcal{P})$.

References

[1] Dirac, P. A. M., *The Principles of Quantum Mechanics* (First Edition), Clarendon Press, Oxford, 1930.

[2] Foulis, D. J., and Randall, C. H., Operational Statistics I: Basis Concepts, J. Math. Physics **13** (1972) pp. 1667–1675.

[3] Foulis, D. J., and Randall, C. H., The Empirical Logic Approach to the Physical Sciences, pp. 230–249 in *Foundations of Quantum Mechanics and Ordered Linear Spaces*, edited by A. Hartkämper and H. Neumann, Lecture Notes in Physics No. 29, Springer Verlag, New York, 1974.

[4] Foulis, D. J., and Randall, C. H., Manuals, Morphisms, and Quantum Mechanics, pp. 105–126 in *Mathematical Foundations of Quantum Theory*, edited by A. R. Marlow, Academic Press, New York, 1978.

[5] Kolmogorov, A. N., *Foundations of the Theory of Probability* (Second Edition), Chelsea Publishing Co., New York, 1956 (German Edition 1933).

[6] Randall, C. H., and Foulis, D. J., Operational Statistics II: Manuals of Operations and their Logics, J. Math. Physics **14** (1973), pp. 1472–1480.

[7] Randall, C. H., and Foulis, D. J., The Operational Approach to Quantum Mechanics, pp. 167–201 in *The Logico-Algebraic Approach to Quantum Mechanics* III: *Physical Theory as a Logico-Algebraic Structure*, edited by C. A. Hooker, D. Reidel Publishing Co., Dordrecht, Holland, 1978.

[8] Randall, C. H., and Foulis, D. J., Operational Statistics and Tensor Products, these Proceedings.
[9] von Neumann, J., *Mathematical Foundations of Quantum Mechanics*, Princeton University Press, Princeton, N. J., 1955.
[10] von Neumann, J., *Continuous Geometry*, Princeton University Press, Princeton, N. J., 1960.
[11] Wright, Ron, Spin Manuals, pp. 177–254 in *Mathematical Foundations of Quantum Theory*, edited by A. R. Marlow, Academic Press, New York, 1978.
[12] Wright, Ron, The State of the Pentagon, pp. 255–274 in *Mathematical Foundations of Quantum Theory*, edited by A. R. Marlow, Academic Press, New York, 1978.

Operational Statistics and Tensor Products

by

C. H. Randall and D. J. Foulis

1. Introduction

This paper is a continuation of [1] and we shall assume that the reader is familiar with its details. Here we assume once and for all that $\mathcal{A}$ and $\mathcal{B}$ are quasimanuals with outcome sets X and Y, respectively. In the present paper we shall discuss the conditioning of regular states on products of $\mathcal{A}$ and $\mathcal{B}$; in particular, we shall be concerned with the tensor product and its relation to orthodox quantum mechanics. In what follows, when we speak of a *state*, we shall always mean a *regular state*, since states that are not regular will not concern us in this paper.

Although $\mathcal{A} \times \mathcal{B}$ is set-theoretically contained in $\overrightarrow{\mathcal{AB}}$, the instructions for executing an operation EF in $\mathcal{A} \times \mathcal{B}$ are not the same as the instructions for executing EF in $\overrightarrow{\mathcal{AB}}$, since in the latter case a definite temporal order is explicit. Nevertheless, there is a natural interpretation morphism $\mathcal{A} \times \mathcal{B} \xrightarrow{h} \mathcal{AB}$ given by $h(xy) = xy$ for all $x \in X$ and all $y \in Y$. Likewise, there are natural interpretation morphisms i, j, and k as shown in the following diagram:

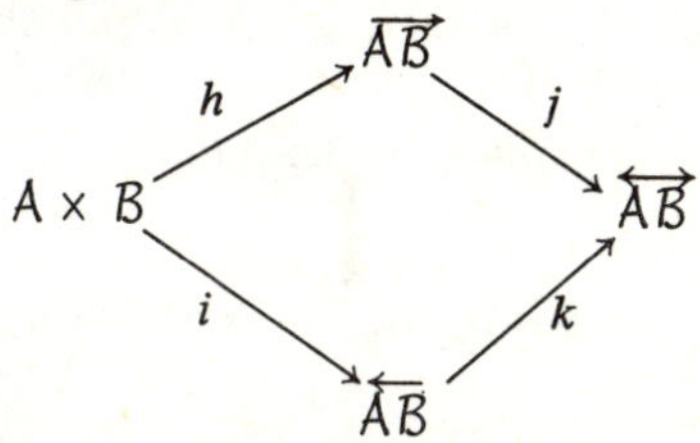

In a similar way, it is operationally more accurate to replace the inclusions of Theorem 9.6 in [1] by the corresponding injective conjugate maps.

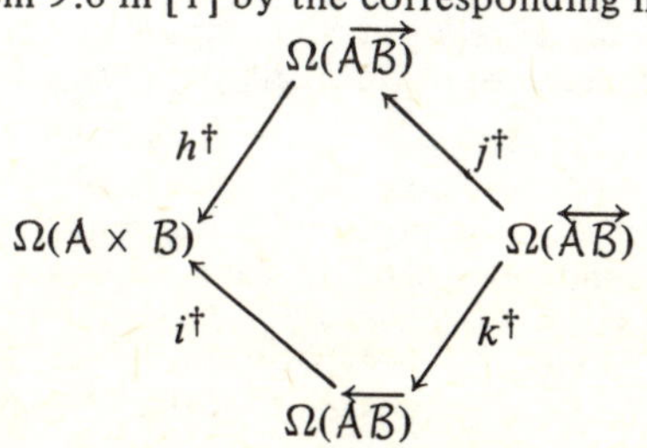

If $\mathcal{A}$ and $\mathcal{B}$ are manuals and $\mathcal{A} \otimes \mathcal{B}$ exists, then we can replace $\overleftrightarrow{\mathcal{A}\mathcal{B}}$ by $\mathcal{A} \otimes \mathcal{B}$ in the diagrams above. In this connection, recall that $\Omega(\mathcal{A} \otimes \mathcal{B}) = \Omega(\overleftrightarrow{\mathcal{A}\mathcal{B}})$.

2. Influence and Conditioning

In this section, let ω be a fixed state in $\Omega(\mathcal{A} \times \mathcal{B})$. If an operation EF in $\mathcal{A} \times \mathcal{B}$ has been executed, what is the probability (in state ω) that $y \in F$ is secured? Evidently, the required probability is given by $\omega(Ey)$. This suggests the following:

2.1 Definition. We define ${}_E\omega \in \Omega(\mathcal{B})$ by ${}_E\omega(y) = \omega(Ey)$ for all $E \in \mathcal{A}$ and all $y \in Y$. We refer to ${}_E\omega$ as the state ω *preconditioned* by the operation E. Similarly, we define $\omega_F \in \Omega(\mathcal{A})$, the state ω *postconditioned* by the operation F in $\mathcal{B}$, by $\omega_F(x) = \omega(xF)$ for all $x \in X$.

If E and G are two $\mathcal{A}$-operations such that ${}_E\omega = {}_G\omega$, we say that E and G have the same *influence* on $\mathcal{B}$. If every pair of $\mathcal{A}$-operations have the same influence on $\mathcal{B}$, we say that $\mathcal{A}$ has *no observable influence* on $\mathcal{B}$. It is convenient to introduce a "do nothing" operation $\{1\}$ for which it is understood that nothing is done and the outcome is automatically 1. If $\{1\} \in \mathcal{A}$, and if $\mathcal{A}$ has no observable influence on $\mathcal{B}$, then we say that $\mathcal{A}$ has *no influence* on $\mathcal{B}$.

2.2 Lemma.

(i) $\omega \in \Omega(\overleftarrow{\mathcal{A}\mathcal{B}})$ if and only if ${}_E\omega = {}_G\omega$ for all $E, G \in \mathcal{A}$.

(ii) $\omega \in \Omega(\overrightarrow{\mathcal{A}\mathcal{B}})$ if and only if $\omega_F = \omega_H$ for all $F, H \in \mathcal{B}$.

As a consequence of Lemma 2.2, if $\mathcal{A}$ and $\mathcal{B}$ are manuals for which $\mathcal{A} \otimes \mathcal{B}$ exists, then $\omega \in \Omega(\mathcal{A} \otimes \mathcal{B})$ if and only if neither $\mathcal{A}$ nor $\mathcal{B}$ has any observable influence on the other (in state ω).

In view of Lemma 2.2, we can make the following definition.

2.3 Definition

(i) If $\omega \in \Omega(\overrightarrow{\mathcal{A}\mathcal{B}})$, $\omega_1 \in \Omega(\mathcal{A})$ is defined by $\omega_1(x) = \omega(xF)$ for any $F \in \mathcal{B}$ and for all $x \in X$.

(ii) If $\omega \in \Omega(\overleftarrow{\mathcal{A}\mathcal{B}})$, $\omega_2 \in \Omega(\mathcal{B})$ is defined by $\omega_2(y) = \omega(Ey)$ for any $E \in \mathcal{A}$ and for all $y \in Y$.

Physicists would refer to ω_1 and ω_2 as *reduced* (or *component*) states for $\mathcal{A}$ and $\mathcal{B}$, respectively, whereas probabilists would call them *marginals.*

Suppose once more that a known operation $EF \in \mathcal{A} \times \mathcal{B}$ has been executed and it is known that the event $A \subseteq E$ occurred. What is the probability that $y \in F$ was secured? Evidently, the required probability is given by $\omega(Ay)/\omega(AF)$. However, suppose that some *unknown* operation $EF \in \mathcal{A} \times \mathcal{B}$ has been executed and that it is known that the event $A \subseteq E$ occurred and that $y \in F$. Here we may not be able to give the probability that y was secured because we cannot account for the possible influence of the unknown operation F – *unless there is no observable influence of such operations on* $\mathcal{A}$. In the latter case, the required probability is clearly given by $\omega(Ay)/\omega_1(A)$. These considerations lead us to the following definition.

2.4 Definition

(i) Let $\omega \in \Omega(\overrightarrow{\mathcal{A}\mathcal{B}})$ and let $A \in \mathcal{E}(\mathcal{A})$ with $\omega_1(A) \neq 0$. Then we define ${}_A\omega \in \Omega(\mathcal{B})$ by ${}_A\omega(y) = \omega(Ay)/\omega_1(A)$ for all $y \in Y$. We refer to ${}_A\omega$ as ω *preconditioned* by A.

(ii) Let $\omega \in \Omega(\overleftarrow{\mathcal{A}\mathcal{B}})$ and let $B \in \mathcal{E}(\mathcal{B})$ with $\omega_2(B) \neq 0$. Then we define $\omega_B \in \Omega(\mathcal{A})$ by $\omega_B(x) = \omega(xB)/\omega_2(B)$ for all $x \in X$. We refer to ω_B as ω *postconditioned* by B.

Naturally, if $x \in X$, ${}_x\omega$ will mean ${}_{\{x\}}\omega$ and, likewise, if $y \in Y$, ω_y will mean $\omega_{\{y\}}$.

As might be expected, the above definitions of conditioning give rise to the following versions of the conventional law of total probability and Bayes theorem.

2.5 Lemma (Law of Total Probability). Let $\omega \in \Omega(\overleftrightarrow{\mathcal{A}\mathcal{B}})$, $E \in \mathcal{A}$, and $F \in \mathcal{B}$. Then:

(i) $\omega_2 = \sum_{x \in E} \omega_1(x)\,{}_x\omega$

and

(ii) $\omega_1 = \sum_{y \in F} \omega_2(y)\,\omega_y$

(where it is understood that terms for which ${}_x\omega$ or ω_y are undefined are to be omitted).

2.6 Theorem (Operational Bayes Theorem). *Let* $\omega \in \Omega(\overleftrightarrow{\mathcal{A}\mathcal{B}})$, $A \in \mathcal{E}(\mathcal{A})$, $B \in \mathcal{E}(\mathcal{B})$, $\omega_1(A) \neq 0$, *and* $\omega_2(B) \neq 0$. *Then*

$$ {}_A\omega(B) = \omega_2(B)\, \frac{\omega_B(A)}{\omega_1(A)} \, . $$

Of course, by replacing the denominator in Theorem 2.6 by the expanded form of $\omega_1(A)$ in part (ii) of Lemma 2.5, we obtain an analogue of the conventional Bayes formula found in elementary textbooks.

In general, we cannot expect different events to pre- or postcondition ω in the same way, even if the two events are perspective. Ron Wright [4], in the process of using empirical logic to study iterated Stern-Gerlach experiments, has given examples of this type of behavior and, as a consequence, has introduced (essentially) the following definition.

2.7 Definition. Let $\omega \in \Omega(\overrightarrow{\mathcal{A}\mathcal{B}})$ and let $A, D \in \mathcal{E}(\mathcal{A})$ with $\omega_1(A) \neq 0$ and $\omega_1(D) \neq 0$. Then, if $p(A) = p(D)$ and ${}_A\omega \neq {}_D\omega$, we say that ω exhibits *interference* between the pair A, D.

Of course, a similar definition applies for $\omega \in \Omega(\overleftarrow{\mathcal{A}\mathcal{B}})$. Roughly speaking, the condition that $\omega \in \Omega(\overleftrightarrow{\mathfrak{A}\mathfrak{B}})$ is equivalent to the condition that ω exhibits no interference whatsoever. More accurately, we have the following lemma.

2.8 Lemma. Let $\omega \in \Omega(\overrightarrow{\mathfrak{A}\mathfrak{B}})$. Then $\omega \in \Omega(\overleftrightarrow{\mathfrak{A}\mathfrak{B}})$ if and only if for every $A, D \in \mathcal{E}(\mathfrak{A}), p(A) = p(D)$ with $\omega_1(A) \neq 0$ implies ${}_A\omega = {}_D\omega$.

As a consequence of the interference between perspective events, it is, in general, impossible to condition a state by a *proposition*. However, because of the preceding lemma, if $\omega \in \Omega(\overleftrightarrow{\mathfrak{A}\mathfrak{B}})$ and if $A \in \mathcal{E}(\mathfrak{A})$ with $\omega_1(A) \neq 0$, we can and do define ${}_{p(A)}\omega = {}_A\omega$. In this connection, it is perhaps pertinent to observe that the well-known Lüder's formula [3] does, in fact, condition Hilbert-space states by propositions (that is, by projection operators on the Hilbert space).

Even if $\omega \in \Omega(\overleftrightarrow{\mathfrak{A}\mathfrak{B}})$, there generally will be events $A, C \in \mathcal{E}(\mathfrak{A})$ and $B \in \mathcal{E}(\mathfrak{B})$ for which ${}_A\omega(B) \neq {}_C\omega(B)$. Although we have seen that such a state ω can not exhibit interference, it may very well exhibit *correlations* between the systems or observers represented by $\mathfrak{A}$ and $\mathfrak{B}$. In this connection, however, notice that a product state $\omega = \alpha\beta$ never exhibits correlations because ${}_A\omega = \beta$ and $\omega_B = \alpha$ whenever they are defined. Conversely, a state that exhibits no correlation must be a product state. In fact, we have the following result.

2.9 Lemma. Let $\omega \in \Omega(\mathfrak{A} \times \mathfrak{B})$. Then the following conditions are equivalent:

(i) ω is a product state.

(ii) $\omega \in \Omega(\overleftrightarrow{\mathfrak{A}\mathfrak{B}})$ and $\omega = \omega_1\omega_2$.

(iii) $\omega \in \Omega(\overrightarrow{\mathfrak{A}\mathfrak{B}})$ and $A, C \in \mathcal{E}(\mathfrak{A})$ with $\omega_1(A) \neq 0$, $\omega_1(C) \neq 0$ implies that ${}_A\omega = {}_C\omega$.

3. Pure States and Conditioning

Recall that a *pure state* for a quasimanual $\mathfrak{A}$ is an element of $\Omega_{\text{ext}}(\mathfrak{A})$, the set of all extreme points of the convex set $\Omega(\mathfrak{A})$.

3.1 Theorem. *Let $\omega \in \Omega(\overrightarrow{\mathfrak{A}\mathfrak{B}})$. Then ω is a pure state if and only if the reduced state ω_1 is pure and every preconditioned state of the form ${}_x\omega$ is pure whenever $\omega_1(x) \neq 0$.*

In spite of the fact that a pure state conditioned by an *outcome* will remain pure, this does not hold in general for conditioning by an *event.* Indeed, if $\omega \in \Omega(\overrightarrow{\mathfrak{A}\mathfrak{B}})$ and $A \in \mathcal{E}(\mathfrak{A})$ with $\omega_1(A) \neq 0$, we have

$$
{}_A\omega = \sum_{x \in A} t(x)\,{}_x\omega
$$

where

$$
\iota(x) = \frac{\omega(x)}{\omega_1(A)} \quad \text{for } x \in A;
$$

hence, since $\sum_{x \in A} t(x) = 1$, it follows that ${}_A\omega$ is a convex combination of the states ${}_x\omega$.

If it is kept in mind that $\Omega(\overleftrightarrow{\mathcal{A}\mathcal{B}}) = \Omega(\mathcal{A} \hat{\otimes} \mathcal{B})$, when $\mathcal{A} \hat{\otimes} \mathcal{B}$ exists, the following lemmas are seen to correspond to well-known results in quantum mechanics [2 pp. 179–182].

3.2 Lemma. Let $\omega \in \Omega(\overleftrightarrow{\mathcal{A}\mathcal{B}})$. Then:

(i) If ω_1 is a pure state, it follows that ω is a product state.

(ii) ω_1 and ω_2 are both pure states if and only if ω is a product of pure states.

Recall that the dispersion free (deterministic) states $\Omega_{df}(\mathcal{A})$ on a quasi-manual $\mathcal{A}$ are always contained in the set $\Omega_{ext}(\mathcal{A})$ of pure states.

3.3 Lemma.

(i) $\Omega_{ext}(\overleftrightarrow{\mathcal{A}\mathcal{B}}) \supseteq \Omega_{ext}(\overrightarrow{\mathcal{A}\mathcal{B}}) \cap \Omega_{ext}(\overleftarrow{\mathcal{A}\mathcal{B}}) = \Omega_{ext}(\mathcal{A})\Omega_{ext}(\mathcal{B})$.

(ii) $\Omega_{df}(\overleftrightarrow{\mathcal{A}\mathcal{B}}) = \Omega_{df}(\mathcal{A})\Omega_{df}(\mathcal{B})$.

Notice, as a consequence of Lemma 3.3, if a pure state on $\overleftrightarrow{\mathcal{A}\mathcal{B}}$ is not a product state, then it cannot be dispersion free.

If all pure states on $\mathcal{A}$ and $\mathcal{B}$ are dispersion free, then the same is true for both operational products $\overrightarrow{\mathcal{A}\mathcal{B}}$ and $\overleftarrow{\mathcal{A}\mathcal{B}}$. However, as the following example shows, this is not necessarily true for $\overleftrightarrow{\mathcal{A}\mathcal{B}}$ (or for $\mathcal{A} \otimes \mathcal{B}$ when it exists).

3.4 Example. The *Janowitz square* is the coherent manual $\mathcal{J} = \{E, F, G, H\}$, where $E = \{e, a, b\}$, $F = \{f, b, c\}$, $G = \{g, c, d\}$, and $H = \{h, d, a\}$. The logic $\Pi(\mathcal{J})$ contains eighteen elements and is the smallest orthomodular partially ordered set that is not a lattice. There are seven pure states on $\mathcal{J}$, all of which are dispersion free, and $\Omega_{df}(\mathcal{J})$ is unital. Therefore [1] there exists a positive Borel interpretation

$$\mathcal{J} \xrightarrow{\Phi} \mathcal{B} = \mathcal{B}(S, \mathcal{M})$$

where $S = \Omega_{df}(\mathcal{J})$ and (in this case) $\mathcal{M}$ is the field of all subsets of S. Furthermore, the conjugate map

$$\Omega(\mathcal{B}) \xrightarrow{\Phi} \Omega(\mathcal{J})$$

is onto, so that all $\mathcal{J}$-states have a classical interpretation. Moreover, $\Phi^\dagger$ maps $\Omega_{df}(\mathcal{B})$ onto $\Omega_{df}(\mathcal{J})$ in a one-to-one fashon. Nevertheless, because $\Phi^\dagger$ is an affine map, $\Omega(\mathcal{B})$ is a simplex, and $\Omega(\mathcal{J})$ is not a simplex, it follows that $\Phi^\dagger$ cannot be one-to-one on all of $\Omega(\mathcal{B})$. It turns out that $\mathcal{J} \otimes \mathcal{J}$ admits a pure state that is not dispersion free.

Thus, as in the previous example, one can obtain the nondeterministic pure states, so characteristic of quantum mechanics, by forming the tensor products of manuals describing essentially classical situations. This state of affairs encourages us to believe that a detailed study of such phenomena may lead to a better understanding of the nondeterministic pure states in quantum mechanics. Here we understand that a premanual $\mathcal{A}$ describes an essentially classical situation if there exists a positive Borel interpretation $\mathcal{A} \xrightarrow{\Phi} \mathcal{B}(S, \mathcal{M})$ whose conjugate $\Phi^\dagger$ maps the Borel states (which are simply the classical probability measures on $\mathcal{M}$) onto $\Omega(\mathcal{A})$, thus providing a classical interpretation for every state on $\mathcal{A}$. In studying tensor products of manuals that describe essentially classical situations, the following lemma is useful.

3.5 Lemma. Let $\mathcal{A}$, $\mathcal{B}$, $\mathcal{C}$, and $\mathcal{D}$ be manuals and suppose that $\mathcal{A} \xrightarrow{\Phi} \mathcal{B}$ and $\mathcal{C} \xrightarrow{\Phi} \mathcal{D}$ be interpretation morphisms. Then there exists a unique interpretation morphism $\mathcal{A} \otimes \mathcal{C} \xrightarrow{\Phi \otimes \theta} \mathcal{B} \otimes \mathcal{D}$ for which $(\Phi \otimes \theta)(ac) = \Phi(a)\theta(c)$. Furthermore, $\Phi \otimes \theta$ is positive if and only if both Φ and θ are positive.

In Example 3.4, we have a positive interpretation morphism $\mathcal{J} \xrightarrow{\Phi} \mathcal{B}(S, \mathcal{M})$ for which $\Phi^\dagger$ is onto. By Lemma 3.5, there is a positive interpretation morphism.

$$\mathcal{J} \otimes \mathcal{J} \xrightarrow{\Phi \otimes \theta} \mathcal{B}(S, \mathcal{M}) \otimes \mathcal{B}(S, \mathcal{M}).$$

As noted, however, $(\Phi \otimes \Phi)^\dagger$ is *not* onto. Thus, $\mathcal{J} \otimes \mathcal{J}$ does not describe an essentially classical situation. Evidently, it will be worthwhile to study the tensor products of Borel manuals. In this connection, the following theorem is of interest.

3.6 Theorem. *Let $(S, \mathcal{M})$ and $(T, \mathcal{N})$ be Borel spaces and let $\mathcal{M} \times \mathcal{N}$ denote the σ-field of subsets of $S \times T$ generated by all measurable rectangles. Then there exists a unique interpretation morphism*

$$\mathcal{B}(S, \mathcal{M}) \otimes \mathcal{B}(T, \mathcal{N}) \xrightarrow{\Phi} \mathcal{B}(S \times T, \mathcal{M} \times \mathcal{N})$$

such that $\Phi(AB) = \{A \times B\}$ for $\phi \neq A \in \mathcal{M}$ and $\phi \neq B \in \mathcal{N}$. Furthermore, the conjugate map $\Phi^\dagger$ is one-to-one and onto.

Since the composition of two positive interpretation morphisms is again one such, an immediate consequence of Lemma 3.5 and Theorem 3.6 is that, if two manuals describe essentially classical situations, then their tensor product admits a positive Borel interpretation; however, as Example 3.4 shows, this tensor product may carry states that are not consistent with this interpretation.

4. Orthodox Quantum Mechanics and Tensor Products

Let us now turn our attention to the special case of orthodox quantum mechanics. Suppose that two quantum mechanical systems are represented in Hilbert spaces $\mathcal{H}_1$ and $\mathcal{H}_2$. Then the union of the two systems is customarily represented in the Hilbert space $\mathcal{H}_1 \otimes \mathcal{H}_2$ the usual Hilbert-space tensor product of H_1 and H_2. In our formalism, the operations – and implicitly, the observables – affiliated with these systems can be represented by the frame manuals [1, Example 7.4] $F_1 = F(H_1)$, $F_2 = F(H_2)$ and $F = F(H_1 \otimes H_2)$. All of these manuals are coherent and carry unital sets of states. Here, the natural question is, how are $F_1 \otimes F_2$ and F related?

It can be shown, to begin with, that, whenever the dimensions of H_1 and H_2 exceed two, it follows that the manual $F_1 \otimes F_2$ is not orthocoherent.

Thus, because F is coherent, the tensor product of the two frame manuals cannot even be isomorphic to the frame manual of the tensor product of the two Hilbert spaces. Notice that the outcome sets for F_1, F_2, and F are the unit spheres S_1, S_2, and S of H_1, H_2, and $H_1 \otimes H_2$, respectively; hence, the outcome set for $F_1 \otimes F_2$ is $S_1 S_2$. The natural map $\Phi : S_1 S_2 \rightarrow S$ given by $\Phi(xy) = x \otimes y$ defines a positive interpretation morphism

$$F_1 \otimes F_2 \xrightarrow{\Phi} F.$$

Because $\Phi((e^{i\theta}x)(e^{-i\theta}y)) = \Phi(xy)$, it follows that Φ is not one-to-one. Thus, it appears that $F_1 \otimes F_2$ carries phase information that is lost in F.

The conjugate map

$$\Omega(F) \xrightarrow{\Phi^\dagger} \Omega(F_1 \otimes F_2).$$

when restricted to pure states, is one-to-one. Indeed, a pure state ω in $\Omega(F)$ can be represented by a unit vector $t \in S$ and the corresponding state $\Phi^\dagger(\omega)$ in $\Omega(F_1 \otimes F_2)$ is evaluated according to the formula $[\Phi^\dagger(\omega)](xy) = |\langle t, x \otimes y\rangle|^2$. Thus, to establish that $\Phi^\dagger$ is one-to-one on such pure states, we must show that the values of $|\langle t, x \otimes y\rangle|$ for $x \in S_1, y \in S_2$ determine t up to a scalar multiple of modulus one.

Recall that each vector $t \in H_1 \otimes H_2$ can be made to correspond to a unique conjugate linear map $\underline{t} : H_2 \rightarrow H_1$ such that

$$\langle t, x \otimes y\rangle = \langle \underline{t}(y), x\rangle$$

for all $x \in H_1$ and $y \in H_2$. (This is done in such a way that $u \otimes v(y) = \langle y, v\rangle u$ for all $u \in H_1$ and $v, y \in H_2$). Thus, the question is reduced to whether $\underline{t}$ is determined (again up to a scalar multiple of modulus one) by the values of $|\langle \underline{t}(y), x\rangle|$ for $x \in S_1$ and $y \in S_2$. Given these values, we can find the linear subspaces $[\underline{t}(y)]^\perp$ for each $y \in S_1$. But this determines the projective transformation corresponding to $\underline{t}$, and thus determines $\underline{t}$ up to a scalar multiple.

In view of the result just obtained, as far as the physical (pure) states are concerned, $F_1 \otimes F_2$ carries as much information as does $F(H_1 \otimes H_2)$. Whether $\Phi^\dagger$ maps $\Omega(F)$ onto $\Omega(F_1 \otimes F_2)$ is at present an open question of some importance. If there are states on $F_1 \otimes F_2$ that have no interpretation in terms of conventional quantum mechanics, one might wonder if any of them are physically realizable.

Since $\mathcal{F}_1 \otimes \mathcal{F}_2$ carries the same information as $\mathcal{F} = \mathcal{F}(\mathcal{H}_1 \otimes \mathcal{H}_2)$, at least as far as the physical (pure) states are concerned, one might justifiably inquire as to precisely how they differ in this respect. The difference, in fact, is that all of the pure states on $\mathcal{F}$ are "observable" (as is the case for the frame manual of any Hilbert space) in the sense that, for any pure state $\omega \in \Omega(\mathcal{F})$, there exists an $\mathcal{F}$-outcome $t \in S$ such that, for every event $A \in \mathcal{E}(\mathcal{F})$.

$$\omega(A) = 1 \quad \text{if and only if} \quad p(\{t\}) \leqslant p(A).$$

However, the state $\Phi^\dagger(\omega)$ may not enjoy "observability" in the manual $\mathcal{F}_1 \otimes \mathcal{F}_2$. Thus, it appears that the coherent interpretation $\mathcal{F}_1 \otimes \mathcal{F}_2 \xrightarrow{\Phi} \mathcal{F}$ appends sufficiently many outcomes to guarantee the "observability" of all physical states.

In summary, we have five essentially different mathematical constructions to represent the union of two quantum mechanical systems, connected by interpretation morphisms as in the following diagram:

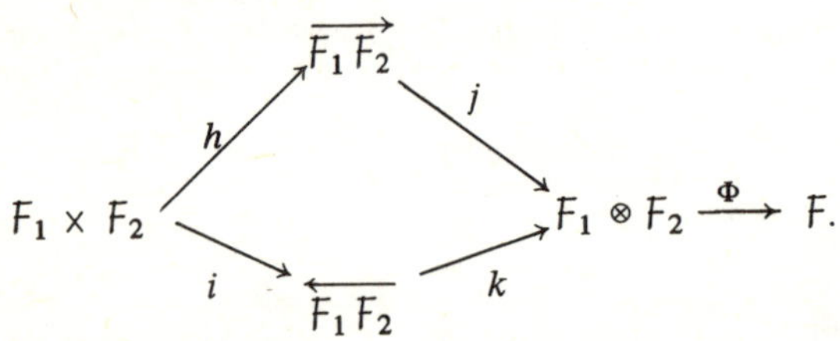

The manual $\mathcal{F}_1 \times \mathcal{F}_2$ admits states that describe bilateral influence, whereas the operational products $\overrightarrow{\mathcal{F}_1 \mathcal{F}_2}$ and $\overleftarrow{\mathcal{F}_1 \mathcal{F}_2}$ only admit states that exhibit unilateral influence. The states on $\mathcal{F}_1 \otimes \mathcal{F}_2$ can describe correlations between the systems, but they can not exhibit influence in either direction. Finally, the physical (pure) states on $\mathcal{F} = \mathcal{F}(\mathcal{H}_1 \otimes \mathcal{H}_2)$ can describe correlations, exhibit no influence, and are "observable."

References

[1] Foulis, D. J., and Randall, C. H., Empirical Logic and Tensor Products, these proceedings.
[2] Jauch, J. M., *Foundations of Quantum Mechanics*, Addison-Wesley Publ. Co., Reading, Mass., 1968.
[3] Lüders, G., Über die Zustandsänderung durch den Meßprozeß, Annalen der Physik, 8 (1951) pp. 322–328.
[4] Wright, Ron, Spin Manuals, pp. 177–254 in *Mathematical Foundations of Quantum Theory*, edited by A. R. Marlow, Academic Press, New York, 1978.

Note added in proof: H. Fischer has recently demonstrated that the conjugate map $\Phi^\dagger$ mentioned on p. 27 is, in fact, one-to-one on all quantum states, pure or mixed.

Some Connections for Manuals of Empirical Logic to Functional Analysis

by

Thurlow A. Cook

In this informal presentation, the theory of manuals of operations is connected with some familiar concepts in functional analysis; namely, base normed and order unit normed spaces. The purpose of this discussion is to present several general open problems which display the interplay of empirical logic with functional analysis. These are mathematical problems with direct physical interpretation.

In this brief talk, I would like to introduce the audience to some of the ways in which linear functional analysis can be used effectively in the study of manuals, as described by Professors Foulis and Randall in this colloquium. The well-known notions of base normed and order unit normed spaces are introduced in order to describe several problems which exist in this theory. Since the spirit of the conference is to give general mathematical ideas with physical import, rather than the most refined mathematical situations, I will pose some elementary problems in the mathematics of manuals which have direct physical appeal. Some of the problems are very general and not well stated, while others may be easy to solve. All, I hope, will give the reader the idea of how broadly applicable this theory is.

Let $\mathcal{A}$ be a manual. For the definition of this basic concept and others like it, one is referred to the papers of Foulis and Randall in this colloquium [4A, B]. Further, we suppose $\Pi(\mathcal{A})$, abbreviated Π, is the logic of $\mathcal{A}$ and Π carries a non-empty set of states $\Omega(\mathcal{A})$, again abbreviated Ω. Since Ω consists of real valued functions on Π, we will always assume (without loss) that Ω is convex and, further, that Ω separates the elements of Π. To keep the discussion as simple as possible and to facilitate the use of linear theory, we will always assume that Ω is, in fact, *strong* over Π; i.e., $p \leqslant q$ in Π iff $\{\omega \in \Omega : \langle \omega, p \rangle = 1\} \subseteq \{\omega \in \Omega : \langle \omega, q \rangle = 1\}$. The notation $\langle \omega, p \rangle$ is used rather than $\omega(p)$ in order to emphasize the dual roles of states and propositions without giving a more fundamental role to either. Let S denote the span of Ω in the real linear space of all real valued functions on Π. Then, Ω acts as a base for the positive cone of S and S can be equipped with two natural norms. These are described as follows: The base norm for S, denoted $\|\cdot\|$ is defined for each $s \in S$ by $\|s\| = \inf\{\alpha + \beta : s = \alpha\omega - \beta\nu, \alpha, \beta \geqslant 0 \text{ in } \mathbb{R}, \omega, \nu \in \Omega\}$. This is easily seen to be the Minkowski functional of the convex hull of $\Omega \cup -\Omega$. The second norm, denoted $\|\cdot\|_Z$, is defined for each $s \in S$ by $\|s\|_Z = \sup\{s(p) - s(p') : p \in \Pi\}$. Then $\|\omega\| = \|\omega\|_Z = 1$ for each $\omega \in \Omega$.

If we now give S the base norm and let S^* be the space of all continuous real valued linear functionals on S, then S^* is the order unit Banach space dual to S. Since Ω is strong over Π, each $p \in \Pi$ represents a unique linear functional on S; in fact, for $p \in \Pi$ we define $\hat{p} \in S^*$ by $\hat{p}(\omega) = \langle \omega, p \rangle$ for each $\omega \in \Omega$. We see with this identification that Π is embedded in S^* in such a way that the ordering of Π as a logic is the ordering of S^* restricted to Π, the null element of Π is the zero of S^*, the unit e of Π is the order unit of S^*, and that for each p in Π, $p' = e - p$ in S^*. Henceforth, we consider $\Pi \subset S^*$ and for simplicity p' is simply denoted p. Since Ω is strong over Π, for each p in Π there exists $\omega \in \Omega$ such that $\langle \omega, p \rangle = 1$ and, therefore, $\|p\| = 1$ in S^*. The unit ball of S^* is $[-e, e] = \{f \in S^* : -e \leqslant f \leqslant e\}$. The set $\Pi \subset [0, e] = \{f \in S^* : 0 \leqslant f \leqslant e\}$ and, for each $s \in S$, $\|s\| = \|s\|_Z$ iff the $w(S^*, S)$-closed convex hull of $\Pi = [0, e]$; see [2]. Again for simplicity, we will always assume these two norms agree.

In many of the traditional presentations of quantum logics the role of a proposition is that of a filter in an "optical bench experiment"; i.e. a proposition $p \in \Pi$ induces a map $\bar{P}$ of the positive cone K of S generated by Ω into K satisfying the following properties:

i) For any $x, y \in K$ and scalar α with $0 \leqslant \alpha \leqslant 1$, $\bar{P}(\alpha x + (1 - \alpha)y) = \alpha\bar{P}(x) + (1 - \alpha)\bar{P}(y)$;

ii) For each $\omega \in \Omega$ there exists some $\nu \in \Omega$ and some scalar α with $0 \leqslant \alpha \leqslant 1$ such that $\bar{P}(\omega) = \alpha\nu$;

and

iii) $\bar{P}^2 = \bar{P}$.

The physical interpretation of these properties is well-known. If each $p \in \Pi$ determines a unique $\bar{P}$, then $\bar{P}$ can easily be extended to a real, positive, linear transformation of S into S such that $\|P\| \leqslant 1$ and $P^2 = P$. These projections are what Alfsen and Shultz [1] call P-projections in their spectral theory of affine function spaces. They show in a von Neumann algebra that each P-projection is of the form $a \to pap$ where p is an idempotent of the algebra [1, p. 95]. One can easily dualize this to obtain that each element of the logic of idempotents induces a unique positive, norm reducing, projection on the space of normal states.

Let us now state the first questions.

Problem 1. What simple hypotheses can be placed on a manual in order that each proposition in the logic induces such a projection on S?

Problem.2. What do all the finite manuals (manuals with finitely many outcomes) look like which have this property?

Let us consider the consequences of having a manual in which each proposition of its logic induces such a transformation. If Ω is strong over Π

and $p \in \Pi$, then $\{\omega \in \Omega : \langle\omega, p\rangle = 1\} = \{\omega \in \Omega : P(\omega) = \omega\}$. Further through the correspondence $p \to \bar{P}$, Π can be embedded in $L(S, S)$, the real linear algebra of bounded linear transformations on S. This allows for a natural interpretation of the product of propositions; namely, a concatenation of experiments. This map of $p \to \bar{P}$ is not obtained in all manuals. Consider the following simple, but rather interesting example; see [4C, p. 1674]. A physical interpretation of the example is given in the aforementioned article. Consider the set of outcomes $X = \{a, b, n, c, d\}$ and the manual A of two operations $\mathsf{A} = \{\{a, b, n\}, \{c, d, n\}\}$. The logic of this manual is as illustrated in fig. 1. It can be shown that the set of states for this logic can be represented as a 3-dimensional pyramid with a square base, and thus the space S of signed states is a 4-dimensional base normed space. In this example the proposition P of the illustrated logic does not induce on S a projection of the type described.

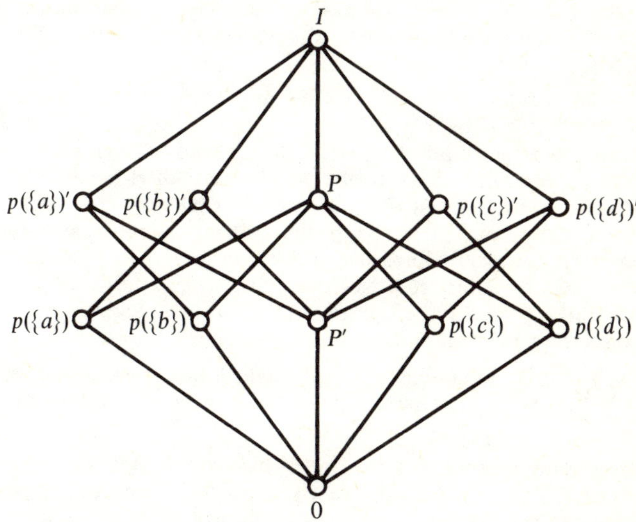

In the papers of Foulis and Randall in this colloquium the notion of an interpretation morphism was introduced. Let us now consider more general morphisms of the manual into itself. By this, I mean the following: If A is our manual and $E(\mathsf{A})$ its set of events, then $\varphi : E(\mathsf{A}) \to E(\mathsf{A})$ is called a *morphism* provided φ satisfies the following two axioms:

i) If $(A_i)_{i \in I}$ is a family of A-events such that $A = \bigcup_i A_i$ is an A-event, then $\varphi(A) = \bigcup_i \varphi(A_i)$.

ii) If A and B are events with $A^\perp \subseteq B^\perp$, then $\varphi(A)^\perp \subseteq \varphi(B)^\perp$.

This notion was introduced in [4D, p. 109]. One can verify the following statements. The composition of morphisms is a morphism. If A and B are weakly perspective events then $\varphi(A)$ and $\varphi(B)$ are also weakly perspective and thus φ induces a mapping, also denoted φ, on II into Π given by $\varphi(p(A)) = p(\varphi(A))$. This map φ induces a map $\varphi^\dagger$ on Ω, called the conjugate of φ, by the formula $\langle \varphi^\dagger \ \omega, p(A) \rangle = \langle \omega, p(\varphi A) \rangle$. It is easily seen that $\varphi^\dagger$ is an affine function on Ω. Let us now refine these notions a little. The morphism φ on $\mathbb{A}$ is called an *automorphism* provided φ is an injective epimorphism and its inverse is also a morphism. An affine bijection of Ω is called a *stochastic symmetry* of Ω and a bijection of Π which preserves all the operations of Π is called a *logical automorphism* or *logical symmetry* of Π. It can be shown that each automorphism of $\mathbb{A}$ induces a logical automorphism of Π and through conjugation a stochastic symmetry of Ω.

Problem 3. Is every stochastic symmetry of Ω induced from an automorphism of $\mathbb{A}$? If the answer to this is no, then find a large class of manuals for which the answer is yes. The answer to this question is yes in the Hilbert space formalism and this is part of the content of Wigner's theorem [5A, p. 144].

Without the use of manuals, it was shown in [3] that the group of automorphisms on Π is isomorphic to the group of symmetries on Ω provided Ω and Π satisfy certain mild properties. In any study of symmetries it will probably be necessary to characterize the elements of Π in some geometric way in S^*. The reason behind this statement is that each stochastic symmetry φ on Ω can be extended to a positive linear isometry, also denoted φ, of S onto itself; thus, its linear adjoint φ^* on S^* is also a positive linear isometry. If one characterizes Π as the set of extreme points of the positive part of the unit ball in S^*, then φ^* restricted to Π is a logical symmetry. This suggests the next problem.

Problem 4. For which manuals with sufficiently many states is it true that Π is exactly the set of extreme points of the positive part of the unit ball of S^*?

In a von Neumann algebra A, Kadison [6] has shown that the idempotents ($P = P^* = P^2$) of this algebra are exactly identified with the extreme boundary of the positive part of the unit ball of A. Rüttimann has solved problem (4) when the manual is finite [8] and the author has obtained a partial result in this direction when the manual is infinite [2]. Alfsen and Shultz obtained a result similar to Kadison's result in their spectral theory memoir [1, p. 68].

With the Hilbert space model of quantum theory before us, the next problem naturally comes to mind. For each bounded operator on a Hilbert space, one can define a bounded linear operator which is the adjont of the given operator. Since the morphisms of a manual induce real linear transformations on S^*, one can ask:

Problem 5. How does one define the adjoint of a morphism on a manual in order that the adjoint is also a morphism?

Probably for most morphisms and manuals this is not possible, but those for which it is possible, should be among the most useful examples. The reader is reminded of the papers of Jauch [5B] and Pool [7] in which the adjoint of a morphism is an essential ingredient.

Let us now connect the last problem with the first.

Problem 6. If the manual $\mathcal{A}$ carries a "sufficiently large" set of morphisms and each morphism admits an adjoint, does each proposition in Π induce a positive linear projection on S; i.e., does each proposition act as a filter on Ω?

The applicability of Baer*-semigroups to this sort of theory was demonstrated by Pool in [7]. In this paper several axioms were given in order to construct an event-state-operation structure in which conditional probability could be interpreted in a non-commutative measure theory. Axiom II.4 on page 127 was introduced in order to be able to define an adjoint of the set of conditioning maps. By a *conditioning morphism* φ on a manual $\mathcal{A}$, we mean a morphism φ with the property: For $\mathcal{A}$-events A and B with $A \perp B$, $\varphi(A) \perp \varphi(B)$; see [4D, p. 110]. The next problem is a very general one.

Problem 7. If the notion of adjoint can be introduced for morphisms on a manual and conditioning is defined as above, can one obtain Pool's theory of conditioning and Baer*-semigroups as a consequence?

In the Hilbert space formalism of quantum mechanics, von Neumann showed that for two propositions (projections) P and Q that $P \wedge Q = \lim_{n \to \infty} (PQ)^n$; see, e.g., [5A, p. 38]. For many manuals Π is a lattice; thus when Π can be embedded in $L(S, S)$ as described earlier, we are led to the next question.

Problem 8. When Π is a lattice and Π is embedded in $L(S, S)$, can one prove that $P \wedge Q = \lim_{n \to \infty} (PQ)^n$?

Let us conclude the discussion with a purely mathematical problem suggested from the theory of von Neumann algebras. It is well-known that a type I factor A has the following property: If (ω_n) is a sequence in Ω which converges to ω in the $w(A_*, A)$-topology, then (ω_n) converges to ω in norm; see [9, p. 21]. It is understood here that A_* is the pre-dual of A generated by the normal states on A. Since the logic of a type I factor is atomic, the following question seems interesting.

Problem 9. If for a manual $\mathcal{A}$ the logic Π is atomic; when is it true that weak-sequential convergence in the positive cone of S implies norm convergence?

Certainly the proof of any theorem of this nature will have to be very different from those used in von Neumann algebra theory. One does not have, a-priori, any useful spectral theory or theory of operator algebras on which to rely. This appears to me to be a very difficult problem – clearly a good place to stop.

References

[1] Alfsen, E. M. and Shultz, F. W., *Non-commutative Spectral Theory for Affine Function Spaces on Convex Sets*, Memoirs A. M. S., 172, 1976.

[2] Cook, T. A., *The Geometry of Generalized Quantum Logics*, International J. of Theo. Phy., (to appear).

[3] Cook, T. A. and Rüttimann, G. T., *Symmetries on Quantum Logics*, (preprint).

[4] Foulis, D. J. and Randall, C. H., (A) *Empirical Logic and Tensor Products*, (these proceedings). (B) *Operational Statistics and Tensor Products*, (these proceedings). (C) *Operational Statistics. I. Basic Concepts*, J. of Math. Phy., **13** (1972), 1667–1675. (D) *Manuals, Morphisms, and Quantum Mechanics*, Mathematical Foundations of Quantum Theory, A. R. Marlow, editor, Proceedings of the conference at Loyola U., New Orleans, June 2–4, 1977, Academic Press (1978), 105–126.

[5] Jauch, J. M., (A) *Foundations of Quantum Mechanics*, Addison-Wesley Pub. Co., 1968. (B) *Scattering Theory in General Quantum Mechanics*, Symposium on Analytic Methods in Mathematical Physics, R. P. Gilbert and R. G. Newton, editors, Conference at U. of Indiana, Bloomington, I., June 2–6, 1968, Gordon and Breach Science Pub. Co. (1970), 185–205.

[6] Kadison, R. V., *Isometries of Operator Algebras*, Ann. of Math. **54** (1951), 325–338.

[7] Pool, J. C. T., *Baer*-Semigroups and the Logics of Quantum Mechanics*, Commun. Math. Phy., 9, (1968), 118–141.

[8] Rüttiman, G. T., *Jordan-Hahn Decomposition of Signed Weights on Finite Orthogonality Spaces*, Commentarii Math. Helv. **52** (1977), 129–144.

[9] Sakai, S., *C*-Algebras and W*-Algebras*, Erg. der Math., **60** (1971), Springer-Verlag.

Detectable Properties and Spectral Quantum Logics

by

Gottfried T. Rüttimann

1. Introduction

We introduce the notions of counter and property as derived concepts within the framework of quantum logics. In the course of our investigations we exhibit a rather important class of quantum logics or orthomodular posets through the requirement that every counter is the expectation functional of some (bounded) observable. These so-called spectral orthomodular posets have many interesting features also as far as the poset of detectable properties is concerned. This study sheds some light on the interrelatedness of the quantum logic and empirical logic approach [5, 6, 10, 15, 16] to Ludwig's approach [12, 13] to the foundations of quantum mechanics both at the mathematical as well as the conceptual level.

2. Preliminaries

We begin with a brief account of the definitions and basic facts pertaining to the subject matter.

A poset $(L, \leqslant)$ with a smallest (0) and a largest element (1) together with a map $p \in L \to p' \in L$ satisfying

(i) if $p \leqslant q'$ then $p \vee q$ exists,

(ii) $(p')' = p$,

(iii) $p \vee p' = 1$,

(iv) $p \leqslant q \Rightarrow q' \leqslant p'$,

(v) $p \leqslant q'$, $p \vee q = 1 \Rightarrow p = q'$

is called *orthomodular*; the elements are called *propositions.* A pair $p, q \in L$, $(L, \leqslant, ')$ an orthomodular poset, is said to be *orthogonal*, denoted $p \perp q$, provided $p \leqslant q'$. A subset $C \subseteq L - \{0\}$ is said to be *orthogonal* if $p, q \in C$ $p \neq q$ implies $p \perp q$. With this definition, $\emptyset \subseteq L$ and singletons $\{p\}$ $(p \neq 0)$ are orthogonal sets. A simple Zorn argument shows that every orthogonal set is contained in a maximal such set. Let $\mathcal{O}(L)$ denote the set of all maximal orthogonal sets and $\mathcal{O}_\sigma(L)$ the set of all maximal orthogonal sets with countably many elements (e.g.: $\{1\}$).

An element $\mu \in \mathbb{R}^L$ is called a *state on L* provided that

$$\mu(p \vee q) = \mu(p) + \mu(q) \text{ if } p \perp q,\ p, q \in L.$$

Note that the states on L form a vector subspace of $\mathbb{R}^L$. A state μ is called *positive* if $\mu(p) \geqslant 0$ for all $p \in L$. The set $K(L)$ of positive states is a cone in $\mathbb{R}^L$. A state μ is said to be *normalized* if $\mu(1) = 1$. A positive normalized state is called *probability state*. We note that the convex set $\Omega(L)$ of probability states is a base for the cone $K(L)$. We denote $V(L) := K(L) - K(L) = \text{lin}\, \Omega(L)$.

Let $E \in \mathcal{O}(L)$ and $\mu \in V(L)$ then the net $(\mu(VC))_{C \in E^f}$, where $(E^f, \subseteq)$ is the set of finite subsets of E directed by set-inclusion, converges in $\mathbb{R}$. Now, a state $\mu \in V(L)$ is called *completely additive*, resp. *σ-additive*, if for every $E \in \mathcal{O}(L)$, resp. $E \in \mathcal{O}_\sigma(L)$, $\mu(1) = \lim\, (\mu(VC))_{C \in E^f}$ holds true. The subspace of $V(L)$ of completely additive, resp. σ-additive, states is denoted by $U_c(L)$, resp. $U_\sigma(L)$. Also, we denote $\Omega_c(L) := U_c(L) \cap \Omega(L)$, $K_c(L) := U_c(L) \cap K(L)$, resp. $\Omega_\sigma(L) := U_\sigma(L) \cap \Omega(L)$, $K_\sigma(L) := U_\sigma(L) \cap K(L)$. Clearly, $K_c(L)$, resp. $K_\sigma(L)$, is a cone in $V(L)$ with $\Omega_c(L)$, resp. $\Omega_\sigma(L)$, as a base. We write $V_c(L) := K_c(L) - K_c(L) = \text{lin}\, \Omega_c(L)$ and $V_\sigma(L) := K_\sigma(L) - K_\sigma(L) = \text{lin}\, \Omega_\sigma(L)$ and remark that $(V(L), \Omega(L))$, resp. $(V_c(L), \Omega_c(L))$, resp. $(V_\sigma(L), \Omega_\sigma(L))$, is a base normed space provided, of course, that $\Omega(L)$, resp. $\Omega_c(L)$, resp. $\Omega_\sigma(L)$, is not empty. For the general theory of base normed spaces and order unit normed spaces see [1, 21] etc. and for a detailed discussion, in various contexts, of the spaces mentioned above refer to [2, 3, 4, 19].

In the sequel we shall be mainly concerned with the base normed space $(V_\sigma(L), \Omega_\sigma(L))$ and its Banachdual (with respect to the base norm) organized to an order unit normed space $(V_\sigma^*(L), \leqslant, e) : f \leqslant g :\Leftrightarrow f(\mu) \leqslant g(\mu)$ for all $\mu \in \Omega_\sigma(L)$, $f, g \in V_\sigma^*(L)$; e is the (unique) linear functional on $V_\sigma(L)$ with $e(\Omega_\sigma(L)) = 1$, an order unit of $(V_\sigma^*(L), \leqslant)$. We set $f' := e - f$ for $f \in V_\sigma^*(L)$. Note that $[0, e]$ $(= \{g \in V_\sigma^*(L) \mid 0 \leqslant g \leqslant e\})$ is closed under the map $f \to f'$. Also, $[-e, e]$ equals the unit ball in $V_\sigma^*(L)$.

With every proposition $p \in L$ we now associate an element of $V_\sigma^*(L)$ as follows: $f_p(\mu) := \mu(p)$. Clearly, $f_p \in [0, e]$ for all $p \in L$, furthermore

(i) $f_0 = 0, f_1 = e$,

(ii) $p \leqslant q \Rightarrow f_p \leqslant f_q$,

(iii) $p \perp q \Rightarrow f_{p \vee q} = f_p + f_q$,

(iv) $f_{p'} = (f_p)'$.

We denote $P := \{f_p \mid p \in L\}$.

As for the conceptual background, we may think of $(L, \leqslant, ')$ as the operational logic of a manual that is considered to be the appropriate "catalogue of questions" for a given empirical system. We also may assume that $\Omega_\sigma(L)$ consists exactly of the regular states, i.e. states induced by complete stochastic models of the manual [5, 6, 16].

3. Counters and Properties

An element $f \in [0, e]$ is called a *counter.* Note that the restriction of a counter to $\Omega_\sigma(L)$ is an affine functional taking on values between 0 and 1. Since every bounded affine functional on $\Omega_\sigma(L)$ admits a unique extension to a base norm continuous linear functional on $V_\sigma(L)$, we might just as well introduce counters as affine functionals on $\Omega_\sigma(L)$ bounded by 0 and 1.

The map $p \in L \to f_p \in [0, e]$ associates with each proposition a counter. We refer to the elements of P as *propositional counters.* Since P is total, i.e. $f(\mu) = 0$ for all $f \in P$ implies that $\mu = 0$, we conclude that $V_\sigma^*(L) = \sigma(V_\sigma^*(L), V_\sigma(L)) - \text{cl}(\text{lin}\, P)$ [11]. Therefore every counter is the weak*-limit of some net of linear combinations of propositional counters.

In this sense, counters are – at least in principle – operationally accessible by the apparatuses described by the manual. Having this in mind, we may interpret the value $f(\mu), f \in [0, e], \mu \in \Omega_\sigma(L)$ as the long-run relative frequency with which the counter f is triggered according to the σ-additive probability state (stochastic model) μ.

Following Mielnik [14], we define a *property* to be a subset $F \subseteq \Omega_\sigma(L)$ such that

for $\mu_1, \mu_2 \in \Omega_\sigma(L)$, $t \in (0, 1)$: $t\mu_1 + (1-t)\mu_2 \in F \Leftrightarrow \mu_1, \mu_2 \in F$.

If $\mu \in \Omega_\sigma(L)$ is the σ-additive probability state describing a given system in the sense of a stochastic model for the manual and $\mu \in F, F$ a property, then we say that *the system has property F.* With $(\mathcal{P}(L), \subseteq)$ we denote the set of properties ordered by set-inclusion, i.e. according to their generality. Clearly, $\emptyset \subseteq \Omega_\sigma(L)$ and $\Omega_\sigma(L)$ are properties, also the set-theoretical intersection of a family of properties is again a property. Therefore $(\mathcal{P}(L), \subseteq)$ is a complete lattice, the infimum of any family of properties being its set-theoretical intersection.

A pair (F, G) of properties is said to be *mutually exclusive*, denoted $F \dagger G$, provided there exists a counter $f \in [0, e]$ with $F \subseteq f^{-1}(1)$ and $G \subseteq f^{-1}(0)$. We say the counter f *separates* the pair of properties (F, G). Note, if (F, G) is separated by f then (G, F) is separated by f', thus the relation $\dagger$ is symmetric. Also, $F \dagger \emptyset$ for all $F \in \mathcal{P}(L)$, e.g. e separates the pair $(F, \emptyset)$. A pair of mutually exclusive properties may be separated by more that one counter.

One easily verifies that for a counter f, $f^{-1}(1) \cap \Omega_\sigma(L)$ and $f^{-1}(0) \cap \Omega_\sigma(L) = (f')^{-1}(1) \cap \Omega_\sigma(L)$ are properties. A property $F \in \mathcal{P}(L)$ is said to be *detectable* provided there exists a counter f such that $F = f^{-1}(1) \cap \Omega_\sigma(L)$; f is referred to as a counter *detecting* the property F. Here too, a detectable property may have more than one counter detecting it. The properties $\emptyset$ and $\Omega_\sigma(L)$ are detectable. With $\mathcal{D}(L)$ we denote the set of detectable properties.

Lemma 3.1: Let, $F, G \in \mathcal{D}(L)$ then $(\mathcal{D}(L), \subseteq) -\inf \{F, G\}$ exists in $(\mathcal{D}(L), \subseteq)$ and equals $F \cap G$.

Proof: There exist $f, g \in [0, e]$ with $F = f^{-1}(1) \cap \Omega_\sigma(L)$, $G = g^{-1}(1) \cap \Omega_\sigma(L)$. Then $1/2\, f + 1/2\, g \in [0, e]$ and therefore $(1/2\, f + 1/2\, g)^{-1}(1) \cap \Omega_\sigma(L)$ is a detectable property. If $1 = (1/2\, f + 1/2\, g)(\mu) = 1/2\, f(\mu) + 1/2\, g(\mu)$ for $\mu \in \Omega_\sigma(L)$ then $f(\mu) = g(\mu) = 1$, so $\mu \in F \cap G$. Conversely, if $\mu \in F \cap G$ then clearly $(1/2\, f + 1/2\, g)(\mu) = 1$. Hence, $(1/2\, f + 1/2\, g)^{-1}(1) \cap \Omega_\sigma(L) = F \cap G$.

A property F is said to be *semi-detectable* provided F is the set-theoretical intersection of some family of detectable properties. Let $\mathcal{S}(L)$ be the set of semi-detectable properties; clearly $\mathcal{D}(L) \subseteq \mathcal{S}(L)$ and the set-theoretical intersection of a family of semi-detectable properties is again semi-detectable. Note that $(\mathcal{P}(L), \subseteq)$ and $(\mathcal{S}(L), \subseteq)$ are complete lattices.

Quite often, propositions are considered as asserting "properties" of the system to be investigated. This can be made more precise as follows: if $p \in L$, we define $\hat{p} = \{\mu \in \Omega_\sigma(L) | \mu(p) = 1\}$, obviously a detectable property detected by the propositional counter f_p. A property F is said to be a *propositional property* provided there exists a proposition $p \in L$ with $\hat{p} = F$. The map $p \in L \to \hat{p} \in \mathcal{D}(L)$ satisfies

(i) $\hat{0} = \emptyset$, $\hat{1} = \Omega_\sigma(L)$,

(ii) $p \leqslant q \Rightarrow \hat{p} \subseteq \hat{q}$,

(iii) $p \perp q \Rightarrow \hat{p} \dagger \hat{q}$.

We say that $\Omega_\sigma(L)$ is *strongly order determining* for $(L, \leqslant, ')$ provided that $\hat{p} \subseteq \hat{q} \Rightarrow p \leqslant q$.

We now focus our interest on detectable properties, specifically on the structure $(\mathcal{D}(L), \subseteq, \dagger)$. Such structures arising from polytopes have been studied in their own right in [17] and measure-theoretic aspects are discussed in [20].

As an illustration let us consider the example outlined on pages 39 and 40.

This example also shows that a detectable property need not be propositional. From a conceptual standpoint of view, however, it is highly desirable to know when the propositions match with the detectable properties, more precisely, when $(L, \leqslant, \perp)$ is isomorphic to $(\mathcal{D}(L), \leqslant, \dagger)$ under the map $p \to \hat{p}$.

Orthomodular poset $(L, \leq, ')$:

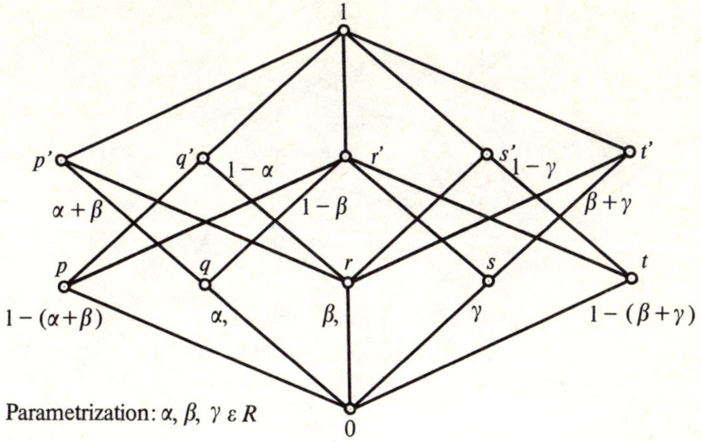

Parametrization: $\alpha, \beta, \gamma \in R$

Convex set of probability measures $\Omega(L) = \Omega_\sigma(L)$:

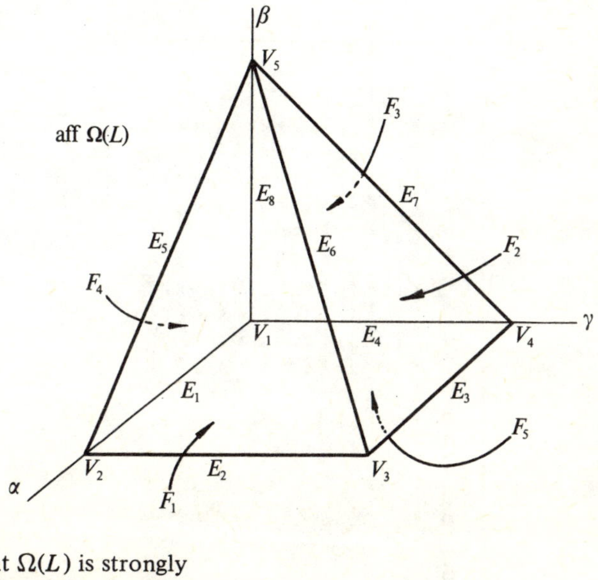

Note that $\Omega(L)$ is strongly order determining for L

Lattice of properties $(\mathcal{P}(L), \subseteq)$:

Note that $\mathcal{P}(L) = \mathcal{D}(L)$

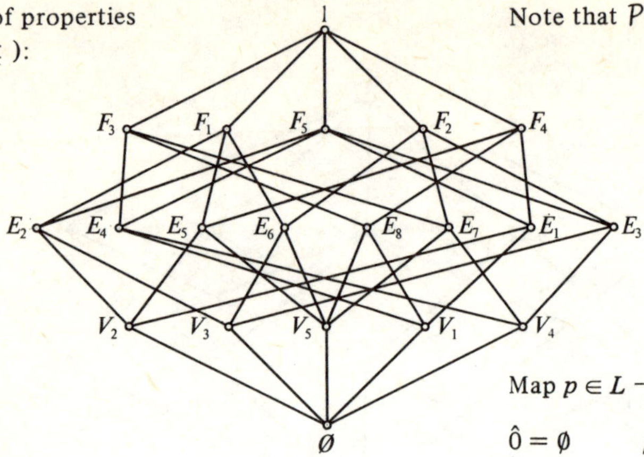

Map $p \in L \to \hat{p} \in \mathcal{D}(L)$:

$\hat{0} = \emptyset$	$\hat{p}' = F_1$
$\hat{r} = V_5$	$\hat{t}' = F_2$
$\hat{t} = E_1$	$\hat{q}' = F_3$
$\hat{q} = E_2$	$\hat{s}' = F_4$
$\hat{s} = E_3$	$\hat{r}' = F_5$
$\hat{p} = E_4$	$\hat{1} = \Omega(L)$

Relation of mutual exclusiveness †:

Property F	$\{G \in \mathcal{P}(L) \mid G \dagger F\}$
$\emptyset$	$\Omega(L)$
V_1	$\emptyset, V_2, V_3, V_4, V_5, E_2, E_3, E_5, E_6, E_7, F_1, F_2$
V_2	$\emptyset, V_1, V_3, V_4, V_5, E_3, E_4, E_6, E_7, E_8, F_2, F_3$
V_3	$\emptyset, V_1, V_2, V_4, V_5, E_1, E_4, E_5, E_7, E_8, F_3, F_4$
V_4	$\emptyset, V_1, V_2, V_3, V_5, E_1, E_2, E_5, E_6, E_8, F_1, F_4$
V_5	$\emptyset, V_1, V_2, V_3, V_4, E_1, E_2, E_3, E_4, F_5$
E_1	$\emptyset, V_3, V_4, V_5, E_3, E_5, E_7, F_2$
E_2	$\emptyset, V_1, V_4, V_5, E_4, E_7, E_8, F_3$
E_3	$\emptyset, V_1, V_2, V_5, E_1, E_5, E_8, F_4$
E_4	$\emptyset, V_2, V_3, V_5, E_2, E_5, E_6, F_1$
E_5	$\emptyset, V_1, V_3, V_4, E_3, E_4$
E_6	$\emptyset, V_1, V_2, V_4, E_1, E_4$
E_7	$\emptyset, V_1, V_2, V_3, E_1, E_2$
E_8	$\emptyset, V_2, V_3, V_4, E_2, E_3$
F_1	$\emptyset, V_1, V_4, E_4$
F_2	$\emptyset, V_1, V_2, E_1$
F_3	$\emptyset, V_2, V_3, E_2$
F_4	$\emptyset, V_3, V_4, E_3$
F_5	$\emptyset, V_5$
$\Omega(L)$	$\emptyset$

4. Spectral Orthomodular Posets

In the sequel we shall always assume that $(L, \leqslant, ')$ is an orthomodular poset such that $\Omega_\sigma(L)$ is strongly order determining for L. With $\mathcal{B}(\mathbb{R})$ we denote the class of Borel sets of $\mathbb{R}$.

A map $x : \mathcal{B}(\mathbb{R}) \to L$ is called *Varadarajan observable of L* (*observable*, for short) provided:

(i) $x(\emptyset) = 0, \quad x(\mathbb{R}) = 1$,

(ii) if $\Delta_1 \cap \Delta_2 = \emptyset, \quad \Delta_1, \Delta_2 \in \mathcal{B}(\mathbb{R})$, then $x(\Delta_1) \perp x(\Delta_2)$,

(iii) for every sequence $(\Delta_i)_{i=1}^{\infty}$ of pairwise disjoint elements in $\mathcal{B}(\mathbb{R})$

$x(\bigcup_{i=1}^{\infty} \Delta_i)$ is the supremum of $\{x(\Delta_i) | \ i = 1, 2, 3, \ldots\}$ in $(L, \leqslant)$.

With $S(L)$ we denote the set of observables of L.

As the definition indicates, we do not assume that $(L, \leqslant, ')$ is σ-orthocomplete but only that the required supremum exists. We may consider the proposition $x(\Delta)$ as asserting that the "measurement" of the observable x yields a real number belonging to the Borel set Δ.

With every proposition $p \in L$ we can associate an observable x_p as follows: $x_p(\Delta) = 0$ if $0 \notin \Delta$ and $1 \notin \Delta$, $x_p(\Delta) = p'$ if $0 \in \Delta$ and $1 \notin \Delta$, $x_p(\Delta) = p$ if $0 \notin \Delta$ and $1 \in \Delta$, $x_p(\Delta) = 1$ if $0 \in \Delta$ and $1 \in \Delta$.

For a Borel function $\alpha : \mathbb{R} \to \mathbb{R}$, i.e. $\alpha^{-1}(\Delta) \in \mathcal{B}(\mathbb{R})$ for all $\Delta \in \mathcal{B}(\mathbb{R})$, and an observable $x \in S(L)$ we define $\alpha(x)(\Delta) := (x \circ \alpha^{-1})(\Delta)\ \Delta \in \mathcal{B}(\mathbb{R})$. Clearly, the map $\Delta \to \alpha(x)(\Delta)$ is an observable.

We define the *spectrum* of an observable $x \in S(L)$ as $s(x) := \cap \{\Delta \subseteq \mathbb{R} \mid \Delta$ closed, $x(\Delta) = 1\}$. One verifies that $x(s(x)) = 1, s(x) \neq \emptyset$ and $s(\alpha(x)) \subseteq cl - \alpha(s(x))$, α a Borel function. An observable x is said to be *positive* if $s(x) \subseteq [0, \infty)$, x is said to be *bounded* if $s(x)$ is bounded, i.e. $s(x)$ is compact. With $S^b(L)$ we denote the set of bounded observables.

Given a probability state $\mu \in \Omega_\sigma(L)$ and an observable $x \in S(L)$ then $\mu \circ x$ is probability measure on $\mathcal{B}(\mathbb{R})$ (in the sense of classical measure theory). Note that $\int \chi_R \ d(\mu \circ x) = \mu \circ x(s(x)) = \mu(x(s(x))) = 1$.

Let us now assume that x is bounded. Then the identity map id : $\mathbb{R} \to \mathbb{R}$ is integrable with respect to $\mu \circ x, \mu \in \Omega_\sigma(L)$. One verifies that

(i) $\inf s(x) \leqslant \int \text{id}\ d(\mu \circ x) \leqslant \sup s(x)$,

(ii) $\inf \mu, \nu \in \Omega_\sigma(L)$ and $t \in [0, 1]$ then

$\int \text{id}\ d((t\mu + (1-t)\nu) \circ x) = \int \text{id}\ d(t(\mu \circ x) + (1-t)(\nu \circ x)) =$
$= t \int \text{id}\ d(\mu \circ x) + (1-t) \int \text{id}\ d(\nu \circ x)$.

From this we conclude that the map $\mu \in \Omega_\sigma(L) \to \int \text{id}\ d(\mu \circ x) \in R$ is a bounded affine functional on $\Omega_\sigma(L)$ and as such it has a unique extension to a base norm continuous linear functional on $V_\sigma(L)$, called the *expectation functional of* $x \in S^b(L)$ and denoted by $E(x)$. The value $E(x)(\mu)$ is called the *expectation of* x *in the state* $\mu \in V_\sigma(L)$. Also note that $\inf s(x)\, e \leqslant E(x) \leqslant \sup s(x)\, e$.

Lemma 4.1: Let $x \in S^b(L)$. Then $\inf s(x) = \inf_{\mu \in \Omega_\sigma(L)} E(x)(\mu)$, $\sup s(x) = \sup_{\mu \in \Omega_\sigma(L)} E(x)(\mu)$, $\|E(x)\| = \max\{\sup s(x), -\inf s(x)\}$.

Proof: We have $\inf s(x) \leqslant E(x)(\mu) \leqslant \sup s(x)$, $\mu \in \Omega_\sigma(L)$. Since $s(x)$ is closed, we conclude that $\inf s(x) \in s(x)$ thus

$$x(\inf s(x) - \epsilon;\ \sup s(x) + \epsilon) \neq 0, \qquad \epsilon > 0.$$

Therefore there exists $\mu \in \Omega_\sigma(L)$ such that $(\mu \circ x)(\inf s(x) - \epsilon; \sup s(x) + \epsilon) = 1$. Then

$$E(x)(\mu) = \int \chi_{(\inf s(x) + \epsilon;\, \inf s(x) + \epsilon)} \operatorname{id} d(\mu \circ x),$$

thus $\inf s(x) \leqslant E(x)(\mu) \leqslant \inf s(x) + \epsilon$, showing that $\inf s(x) = \inf_{\mu \in \Omega_\sigma(L)} E(x)(\mu)$. Similarly, $\sup s(x) = \sup_{\mu \in \Omega_\sigma(L)} E(x)(\mu)$. The final assertion follows from $\sup_{\mu \in \Omega_\sigma(L)} |E(x)(\mu)| = \sup_{\mu \in \operatorname{con}(\Omega_\sigma(L) \cup -\Omega_\sigma(L))} E(x)(\mu) = \|E(x)\|$.

Lemma 4.2: Let $x \in S(L)$ with $s(x) \subseteq [0, 1]$. Then

(i) $E(x)^{-1}(0) \cap \Omega_\sigma(L) = \widehat{x\{0\}}$,

(ii) $E(x)^{-1}(1) \cap \Omega_\sigma(L) = \widehat{x\{1\}}$.

Proof: (i): Let $E(x)(\mu) = 0$, $\mu \in \Omega_\sigma(L)$. Then $0 = \int \operatorname{id} d(\mu \circ x) = \int (\chi_{\{0\}} + \chi_{(0,1]} \operatorname{id} d(\mu \circ x) = \int \chi_{(0,1]} \operatorname{id} d(\mu \circ x)$. For $t \in (0, 1]$ $\chi_{(0,1]} \operatorname{id} > 0$, we conclude that $(\mu \circ x)(0, 1] = \mu(x(0, 1]) = 0$. Thus $\mu(x\{0\}) = 1$ since $(\mu \circ x)[0, 1] = 1$. Therefore $\mu \in \widehat{x\{0\}}$. The converse is easily shown.

(ii): If $E(x)(\mu) = 1$, $\mu \in \Omega_\sigma(L)$ then $E(x)'(\mu) = \int (1 - t)\, d(\mu \circ x) = 0$. Now use similar arguments as in (i).

It follows from the foregoing observations that $E(x)$ $(x \in S^b(L))$ is a counter if and only if $s(x) \subseteq [0, 1]$. However, in general there are counters that do not arise in this manner. We get hold of a rather important class of orthomodular posets (with a strongly order determining $\Omega_\sigma(L)$) through the following definition.

The orthomodular poset $(L, \leqslant, ')$ is said to be *spectral* provided every counter is the expectation functional of some bounded observable. Without proof we note that the usual orthomodular poset of projections of a Hilbert space of dimension greater than 2 is spectral.

More insight is gained through

Theorem 4.3: *The following are mutually equivalent:*

(i) *$(L, \leqslant, ')$ is spectral,*

(ii) *to every $f \in V_\sigma^*(L)$ there exists $x \in S^b(L)$ such that $f = E(x)$,*

(iii) a) *if $x, y \in S^b(L)$ then there exists an observable $z \in S^b(L)$ such that $E(z)(\mu) = E(x)(\mu) + E(y)(\mu)$ for all $\mu \in \Omega_\sigma(L)$ and*

b) *if $(x_\delta)_\delta$ is a net of positive bounded observables such that* $\lim(E(x_\delta)(\mu))_\delta$ *exists for all $\mu \in \Omega_\sigma(L)$ and is uniformly bounded in $\mu \in \Omega_\sigma(L)$ then there exists an observable $z \in S^b(L)$ such that $E(z)(\mu) = \lim\,(E(x_\delta)(\mu))_\delta$ for $\mu \in \Omega_\sigma(L)$.*

Proof: (i) ⇒ (ii): Let $g \in V_\sigma^*(L)$, we may assume $g \neq 0 = E(x_0)$. Then $1/2\,(g/\|g\| + e) \in [0, e]$. Now let $x \in S^b(L)$ be such that $E(x) = 1/2\,(g/\|g\| + e)$. The Borel function $\alpha(t) := 2\|g\|t - \|g\|$ is bounded on $s(x)$, hence $s(\alpha(x))$ is bounded showing that $\alpha(x) \in S^b(L)$. We now get for all $\mu \in \Omega_\sigma(L)$, using the "transformation theorem of integration" [9, p. 163], $E(\alpha(x))(\mu) = \int \mathrm{id}\, d(\mu \circ (x \circ \alpha^{-1})) = \int \mathrm{id}\, d((\mu \circ x) \circ \alpha^{-1}) = \int \alpha(t)\, d(\mu \circ x) = \int (2\|g\|t - \|g\|)\, d(\mu \circ x) = 2\|g\| \int \mathrm{id}\, d(\mu \circ x) - \|g\| = g(\mu)$. Therefore $E(\alpha(x)) = g$.

(ii) ⇒ (i): This is immediate.

(ii) ⇒ (iii): Clearly, condition a) holds since $V_\sigma^*(L)$ is a linear space. If $(E(x_\delta)(\mu))_\delta$ converges for all $\mu \in \Omega_\sigma(L)$ then it converges for all $\mu \in V_\sigma(L)$. Also the map $\mu \in V_\sigma(L) \to f(\mu) := \lim\limits_\delta E(x_\delta)(\mu) \in \mathbb{R}$ is a linear functional on $V_\sigma(L)$. Since $0 \leqslant \inf s(x_\delta) \leqslant E(x_\delta)(\mu)$ and $f(\mu) \leqslant k \in \mathbb{R}$ for all $\mu \in \Omega_\sigma(L)$, we conclude that $f \in V_\sigma^*(L)$. Now $f = E(x)$ for some $x \in S^b(L)$ and assertion (b) follows.

(iii) ⇒ (ii): First observe that $E := \{E(x) \mid x \in S^b(L)\}$ is a linear subspace of $V_\sigma^*(L)$ since $sE(x) = E(\alpha(x)), \alpha(t) = st\,(s \in \mathbb{R})$ and to $x, y \in S^b(L)$ there exists $z \in S^b(L)$ with $E(x) + E(y) = E(z)$. Also $P \subseteq E$, since $E(x_p) = f_p$ for all $p \in L$, and therefore $\mathrm{pos}\, P \subseteq E$, $\mathrm{pos}\, P$ being the positive hull of P in $V_\sigma^*(L)$.

Let $(f_\delta)_\delta$ be a net in $\mathrm{pos}\, P$ that converges to $f \in V_\sigma^*(L)$ in the $\sigma(V_\sigma^*(L), V_\sigma(L))$-topology. To every f_δ there exists a positive observable $x_\delta \in S^b(L)$ (see lemma 4.1) with $f_\delta = E(x_\delta)$. Then clearly $\lim\limits_\delta E(x_\delta)(\mu) = f(\mu) \leqslant \|f\|$ for all $\mu \in \Omega_\sigma(L)$ since $0 \leqslant f$. By condition b) there exists now an $x \in S^b(L)$ such that $E(x) = f$. Showing that $\sigma(V_\sigma^*(L), V_\sigma(L)) - \mathrm{cl}\,(\mathrm{pos\,P}) \subseteq E$.

Consider now the dual pair $(V_\sigma^*(L), V_\sigma(L))$ with scalar product $(f, \mu) := f(\mu)$. By the bipolar theorem [11], we conclude that $(\mathrm{pos}\,P)^{00} = \sigma(V_\sigma^*(L), V_\sigma(L)) - \mathrm{cl}(\mathrm{pos}\ \mathrm{P})$, since $\mathrm{pos}\,P$ is convex and $0 \in \mathrm{pos}\,P$. Since $s(\mathrm{pos}\,P) = \mathrm{pos}\,P$ for $s > 0$ we have $(\mathrm{pos}\,P)^0 = \{\mu \in V_\sigma(L) \mid f(\mu) \leqslant 1 \text{ for all } f \in \mathrm{pos}\,P\} = \{\mu \in V_\sigma(L) \mid f(\mu) \leqslant 0 \text{ for all } f \in \mathrm{pos}\,P\}$. Then $(\mathrm{pos}\,P)^0 = \{\mu \in V_\sigma(L) | \mu(p) \leqslant 0, \text{ all } p \in L\} = -K_\sigma(L)$ and finally $(\mathrm{pos}\,P)^{00} = (-K_\sigma)^0 = \{f \in V_\sigma^*(L) \mid f \geqslant 0\}$. Combining with the above results, we get $\{f \in V_\sigma^*(L) \mid f \geqslant 0\} \subseteq E$. Since the set on the left-hand side is a generating cone we get $V_\sigma^*(L) \subseteq E$.

Spectrality implies that $(L, \leqslant, \perp)$ is isomorphic to the structure $(\mathcal{D}(L), \subseteq, \dagger)$:

Theorem 4.4: *Let $(L, \leqslant, ')$ be spectral. Then the map $p \in L \to \hat{p} \in \mathcal{D}(L)$ is surjective, $p \leqslant q \Leftrightarrow \hat{p} \subseteq \hat{q}$ and $p \perp q \Leftrightarrow \hat{p} \dagger \hat{q}$.*

Proof: Let $F \in \mathcal{D}(L)$ then there exists $f \in [0, e]$ such that $f^{-1}(1) \cap \Omega_\sigma(L) = F$. $(L, \leqslant, ')$ being spectral, there exists $x \in S^b(L)$ with $f = E(x)$. Since $s(x) \subseteq [0, 1]$ we conclude, by lemma 4.2, that $F = \hat{x}\{1\}$. The second assertion is immediate. Let $\hat{p} \dagger \hat{q}$. Then there is an $x \in S^b(L)$ with $E(x) \in [0, e]$, $\hat{p} \subseteq E(x)^{-1}(1) \cap \Omega_\sigma(L)$ and $\hat{q} \subseteq E(x)^{-1}(0) \cap \Omega_\sigma(L)$. Again by lemma 4.2, $\hat{p} \subseteq x\{1\}, \hat{q} \subseteq x\{0\}$ thus $p \leqslant x\{1\},\ q \leqslant x\{0\}$. Therefore, $p \perp q$ since $\{1\} \cap \{0\} = \emptyset$.

Lemma 4.5: *If $(L, \leqslant, ')$ is spectral then $(L, \leqslant)$ is a lattice and for all $\mu \in \Omega_\sigma(L)$ $\mu(p) = \mu(q) = 1 \Rightarrow \mu(p \wedge q) = 1$.*

Proof: Let $p, q \in L$. By lemma 3.1 $(\mathcal{D}(L), \subseteq) - \inf\{\hat{p}, \hat{q}\}$ exists and equals $p \cap q$. Then clearly $p \wedge q$ exists, by the foregoing theorem, hence $(L, \subseteq)$ is a lattice. Also $(\hat{p} \wedge \hat{q}) = \hat{p} \cap \hat{q}$.

Theorem 4.3 shows that a spectral orthomodular poset satisfies the "existence condition" in the sense of S. P. Gudder, therefore the above result may be considered as an application of [7, lemma 1.4]. As observed in [8], combining these results with a theorem by the author [18, theorem 4.3] we obtain the

Theorem 4.6: *Let L be finite. Then $(L, \leqslant, ')$ is spectral if and only if $(L, \leqslant)$ is a Boolean lattice.*

We shall make use of the following technical

Lemma 4.7: Every monotonically decreasing net $(f_\delta)_\delta$ in $[0, e]$ has a $\sigma(V_\sigma^*(L), V_\sigma(L))$-limit f in $[0, e]$. Also $f \leqslant f_\delta$.

Proof: First note that $[0, e]$ is $\sigma(V_\sigma^*(L), V_\sigma(L))$-compact since it is the image of the unit ball in $V_\sigma^*(L)$ under the $\sigma(V_\sigma^*(L), V_\sigma(L))$-homeomorphism $g \to (g + e)/2$. Therefore there exists a subnet $(f_{\delta'})_{\delta'}$ and $f \in [0, e]$ such that $f = \sigma(V_\sigma^*(L), V_\sigma(L)) - \lim f_{\delta'}$.

To this end, for every $\mu \in \Omega_\sigma(L)$, $(f_\delta(\mu))_\delta$ is a monotonically decreasing net in $[0, 1] \subseteq R$ and therefore converges. The subnet $(f_{\delta'}(\mu))_{\delta'}$ converges to $f(\mu)$, $\mu \in \Omega_\sigma(L)$. Hence $f_\delta(\mu) \to f(\mu)$ and $f(\mu) \leqslant f_\delta(\mu)$ for all $\mu \in \Omega_\sigma(L)$ and we are done.

Theorem 4.8: *Let* $(L, \leqslant, ')$ *be spectral. Then*

(i) $(L, \leqslant)$ *is a complete lattice,*

(ii) *for any family* $(p_i)_{i\in I}$ *in* L *and* $\mu \in \Omega_\sigma(L)$: *if* $\mu(p_i) = 1$ for all $i \in I$ *then* $\mu(\bigwedge_{i\in I} p_i) = 1$.

Proof: Let D be a non-empty subset of L and denote with $(D^f, \subseteq)$ the set of finite subsets of D directed by set-inclusion. By lemma 4.5, $\bigwedge C$ exists for $C \in D^f$. Then $(f_{\bigwedge C})_{C\in D^f}$ is a monotonically decreasing net in $[0, e]$ and, by the foregoing lemma, there exists $f \in [0, e]$ with $f \leqslant f_C$, $C \in D^f$, and $f = (V_\sigma^*(L), V_\sigma(L)) - \lim(f_{\wedge C})_{C\in D^f}$. Then in particular $f \leqslant f_p$ for all $p \in D$, thus $f^{-1}(1) \cap \Omega_\sigma(L) \subseteq \bigcap_{p\in D} \hat{p}$. Conversely, let $\mu \in \bigcap_{p\in D} \hat{p}$ then $f(\mu) = (\sigma(V_\sigma^*(L), V_\sigma(L) - \lim\,(f_{\wedge C})_{C\in D^f})\,(\mu) =$ $\lim(\mu(\bigwedge C))_{C\in D^f} = 1$, by the second statement in lemma 4.5. Consequently, $f^{-1}(1) \cap \Omega_\sigma(L) = \bigcap_{p\in D} \hat{p}$. Since $f^{-1}(1) \cap \Omega_\sigma(L) \in \mathcal{D}(L)$ there exists, by theorem 4.4 a proposition $q \in L$ with $\hat{q} = \bigcap_{p\in D} \hat{p}$. Then $\bigwedge_{p\in D} p$ exists and equals q again by theorem 4.4. The remainder follows immediately.

Corollary 4.9: *If* $(L, \leqslant, ')$ is spectral then $(\mathcal{D}(L), \subseteq)$ a complete lattice and every semi-detectable property is a detectable property.

Theorem 4.10: *If* $(L, \leqslant, ')$ *is spectral then every* σ*-additive probability state is completely additive.*

Proof: The assertion holds true in any complete orthomodular lattice satisfying (ii) of theorem 4.8:

Let $E \in \mathcal{O}$ and $\mu \in \Omega_\sigma(L)$. Clearly, the net $(\mu(\bigvee C))_{C\in E^f}$ converges and $\mu(\bigvee C) = \sum_{p\in C} \mu(p)$ for $C \in E^f$. It is a well-known fact from real analysis that there exists a sequence $n \in N \to p_n \in E$ such that $\mu(p) = 0$ for all $p \in D := E - \{p_n \mid n \in N\}$ and $\lim\,(\sum_{p\in C} \mu(p))_{C\in E^f} = \sum_{n=1}^{\infty} \mu(p_n)$.

Now, since E is a maximal orthogonal set, we get $1 = \bigvee_{i=1}^{\infty} p_i \vee \bigvee D$ and therefore $\mu(1) = \mu(\bigvee_{n=1}^{\infty} p_n) + \mu(\bigvee D) = \sum_{n=1}^{\infty} \mu(p_n) + 0 = \lim (\mu(\bigvee C))_{C \in E^f}$.

5. References

[1] E. M. Alfsen: *Compact Convex Sets and Boundary Integrals*, Springer Verlag, Berlin, 1971.
[2] T. A. Cook & G. T. Rüttimann: "Symmetries on Quantum Logics", Preprint University of Massachusetts.
[3] T. A. Cook: "The Geometry of Generalized Quantum Logics". International Journal of Theoretical Physics. 17, 941 (1978).
[4] H. R. Fischer & G. T. Rüttimann: "The Geometry of the State space", in *Mathematical Foundations of Quantum Theories*, editor A. R. Marlow, Academic Press, New York, 1978, pp. 153–176.
[5] D. J. Foulis & C. H. Randall: "Operational Statistics I", Journal of Mathematical Physics 13, 1667 (1972).
[6] D. J. Foulis & C. H. Randall: this volume.
[7] S. P. Gudder: "Uniqueness and Existence Properties of Bounded Observables", Pacific Journal of Mathematics 19, 81 (1966).
[8] S. P. Gudder: "Some Unsolved Problems in Quantum Logics", in *Mathematical Foundations of Quantum Theories*, editor A. R. Marlow, Academic Press, New York, 1978, pp. 87–103.
[9] P. R. Halmos: *Measure Theory*, Van Nostrand Reinhold, New York, 1950.
[10] J. M. Jauch: *Foundations of Quantum Mechanics*, Addison-Wesley Publishing Company, Reading Mass., 1968.
[11] G. Köthe: *Topologische lineare Räume*, Springer Verlag, Berlin, 1960.
[12] G. Ludwig: *Deutung des Begriffs „physikalische Theorie" und axiomatische Grundlegung der Hilbertraumstruktur der Quantenmechanik durch Hauptsätze des Messens*, Springer Verlag, Berlin, 1970.
[13] G. Ludwig: this volume.
[14] B. Mielnik: "Generalized Quantum Mechanics", Communications in Mathematical Physics 37, 221 (1974).
[15] C. Piron: *Foundations of Quantum Physics*, Benjamin Inc., Reading Mass., 1976.
[16] C. H. Randall & D. J. Foulis: "Operational Statistics II", Journal of Mathematical Physics 14, 1472 (1973).
[17] G. T. Rüttimann: "Stable Faces of a Polytope", Bulletin of the American Mathematical Society 82, 314 (1976).
[18] G. T. Rüttimann: "Jauch-Piron States", Journal of Mathematical Physics 18, 189 (1977).
[19] G. T. Rüttimann: "Jordan-Hahn Decomposition of Signed Weights on Finite Orthogonality Spaces", Commentarii Mathematici Helvetici 52, 129 (1977).

[20] G. T. Rüttimann: "Non-commutative Measure Theory", Habilitationsschrift Universität Bern, 1979.
[21] H. H. Schaefer: *Topological Vector Spaces*, Springer Verlag, New York, 1971.
[22] V. S. Varadarajan: *Geometry of Quantum Theory*, Van Nostrand Company, London, 1968.

An Axiomatic Basis of Quantum Mechanics

by

G. Ludwig

The difficulties arising in the usual interpretation of quantum mechanics are caused by using "states" and "observables" as basic notions, notions which have to be explained by quantum mechanics itself. We will not describe these difficulties in the usual interpretation of quantum mechanics. Instead of this we will present an interpretation starting *only* with the well-known description of macroscopic systems, i.e. a classical, objective description.

Microsystems, like single electrons, single atoms, single molecules and so on, "appear" in experiments quite indirectly by processes occurring on macro-systems, processes which can be stated as happening according to well-known classical theories or direct seeing or hearing. Every experiment with single microsystems is, in any case, an experiment using macrosystems. Microsystems are introduced to *explain* what has happened on the macrosystems.

This procedure, starting a theory from those facts which can be *described* without the considered theory itself, is called the construction of an *axiomatic basis* for a physical theory [3].

The aim of the construction of such an axiomatic basis is to improve the understanding of the theory and – more essentially – to get ontological propositions concerning a new reality region, discovered by the theory (see [1]). By this method the discovery of the microsystems can be described [2], [7]. In this contribution we will take for granted the existence of the microsystems. We assume that the preparation and registration procedures of these microsystems may be described by well-known classical pretheories. We do not assume that we can *explain* the interaction of the microsystems with the macroscopic apparatuses without quantum mechanics. However, we can *describe* the effect processes produced by the microsystems on the apparatuses without knowing a theory of the interaction between microsystems and the macroscopic apparatuses.

To give a mathematical description of the preparation and registration procedures we will introduce a general structure to describe statistics. Since rather detailed descriptions of these structures may by found in [1] §§ 11, 12; [3]; [4]; [5] XIII, it is sufficient to give a short survey here.

1. Statistics

The most general description of statistics in physics is gained by introducing the concept of statistical selection procedures. The basic set M for this structure will be used as a set of labels for those facts, which shall be selected according to various procedures. We introduce the structure $\mathcal{S}$ of species selection procedures as a set $\mathcal{S} \subset \mathcal{P}(M)$ which satisfies the axioms:

AS 1.1 For $a, b \in \mathcal{S}, a \subset b$ holds, $b \setminus a \in \mathcal{S}$;

AS 1.2 For $a, b \in \mathcal{S}$ holds, $a \cap b \in \mathcal{S}$.

The elements of $\mathcal{S}$ are called selection procedures. $x \in a$ (with $a \in \mathcal{S}$) is the translation of the fact that x is selected according to the procedure a.

Statistical dependence of the selection procedures are described by a probability function

$\lambda : \mathcal{T} = \{(a, b) : a, b \in \mathcal{S}, b \subset a, a \neq \phi\} \to [0, 1]$

satisfying the axioms

AS 2.1 If $a_1, a_2 \in \mathcal{S}, a_1 \cap a_2 = \phi, a_1 \cup a_2 \in \mathcal{S}$ then $\lambda(a_1 \cup a_2, a_1) + \lambda(a_1 \cup a_2, a_2) = 1$;

AS 2.2 If $a_1, a_2, a_3 \in \mathcal{S}, a_1 \supset a_2 \supset a_3, a_2 \neq \phi$ then $\lambda(a_1, a_3) = \lambda(a_1, a_2)\, \lambda(a_2, a_3)$;

AS 2.3 If $a_1, a_2 \in \mathcal{S}, a_2 \subset a_1; a_2 \neq \phi$ then $\lambda(a_1, a_2) \neq 0$.

$\lambda(a, b)$ is called the probability of b relative to a.

2. Preparation and registration procedures

Experiments with microsystems consist of preparing these systems and afterwards measuring them. We will only take into account so-called digital measurements, which is in fact no restriction of generality. The digital measurements will be described by registration procedures.

Let M be the basic set the elements of which are used to label the microsystems. The preparation procedures are described by a set $Q \subset \mathcal{P}(M)$ for which is assumed

APS 1 Q is a structure of species statistical selection procedure.

The mathematical relation $x \in a$ (with $a \in Q$) is the translation of the statement: x is prepared by the preparation procedure a;

We shall denote the probability function for Q by λ_Q.

It is more difficult to present a mathematical formulation of the registration processes. A registration process is characterized by two steps:

(1) Construction and application of the registration apparatus

(2) Selection according to the changes which occur (or do not occur) at the registration apparatus.

Accordingly, we shall define an additional structure on M by means of two subsets of $\mathcal{P}(M)$: $\mathcal{R}$ and $\mathcal{R}_0$. For $\mathcal{R}$ and $\mathcal{R}_0$ we require the following axioms be satisfied:

APS 2 $\mathcal{R}$ is a structure of species selection procedure

APS 3 $\mathcal{R}_0$ is a structure of species statistical selection procedure

APS 4.1 $\mathcal{R}_0 \subset \mathcal{R}$

APS 4.2 To each $b \in \mathcal{R}$ there is a $b_0 \in \mathcal{R}_0$ such that $b \subset b_0$.

The mathematical relation $x \in b_0$ may be the translation of the statement: For the registration of x the method b_0 (the apparatus characterized by b_0) is applied. The mathematical relation $x \in b \subset b_0$ may be the translation of the statement: registering x the change b has occurred at the applied apparatus b_0.

The elements of $\mathcal{R}_0$ are called registration methods, the elements of $\mathcal{R}$ are called registration procedures. The probability function for $\mathcal{R}_0$ will be denoted by $\lambda_{\mathcal{R}_0}$.

We shall now consider the question which combinations of $a \in \mathcal{Q}$ and $b_0 \in \mathcal{R}_0$ are physically meaningful. This problem is by no means trivial. In order to formulate this problem in the mathematical language we introduce the following definition:

Let $a \in \mathcal{Q}, b_0 \in \mathcal{R}_0, a \neq \phi, b_0 \neq \phi$; we say that a and b_0 *may be combined* providing that if $\tilde{a} \in \mathcal{Q}, \tilde{b}_0 \in \mathcal{R}_0, \phi \neq \tilde{a} \subset a, \phi \neq \tilde{b}_0 \subset b_0$ we find that $\tilde{a} \cap \tilde{b}_0 \neq \phi$.

Let $C = \{(a, b_0) : a \in \mathcal{Q}, b_0 \in \mathcal{R}_0$ and a, b_0 may be combined$\}$.

We note the following assumptions concerning C:

APS 5.1.1 To each $a \in \mathcal{Q}, a \neq \phi$ there exists a $b_0 \in \mathcal{R}_0$ such that $(a, b_0) \in C$

APS 5.2 To each $b \in \mathcal{R}, b \neq \phi$ there exists at least one $a \in \mathcal{Q}$ and a $b_0 \in \mathcal{R}_0$ which satisfies $b \subset b_0, (a, b_0) \in C$ and $a \cap b \neq \phi$.

Axiom APS 5.2 is stronger than APS 4.2.

We shall now consider the mathematical formulation of the statistical dependence of the combined selection procedures. First we define

$\theta = \{c : c = a \cap b, a \in \mathcal{Q}, b \in \mathcal{R}$ and for $a \neq \phi$ there exists a $b_0 \in \mathcal{R}_0$ such that $b \subset b_0$ and $(a, b_0) \in C\}$.

An element $a \cap b$ of θ represents the microsystems which are prepared according to a and registered according to b.

Let $\mathcal{S}$ denote the smallest set of selection procedures such that $\theta \subset \mathcal{S}$.

By the following requirement is expressed that the combination of preparation and registration procedures lead to reproducible frequencies.

APS 6 $\mathcal{S}$ is a structure of species statistical selection procedures.

Let $\lambda_{\mathcal{S}}$ denote the probability function for $\mathcal{S}$.

On the basis of their physical interpretation λ_Q, λ_{R_0} and λ_S can't be mutually independent. We assume:

APS 7 Let $a_1, a_2 \in Q$, $b_{10}, b_{20} \in R_0$ where $a_2 \subset a_1$, $b_{20} \subset b_{10}$ and $(a_1, b_{10}) \in C$. Then

APS 7.1 $\lambda_S(a_1 \cap b_{10}, a_2 \cap b_{10}) = \lambda_Q(a_1, a_2)$;

APS 7.2 $\lambda_S(a_1 \cap b_{10}, a_1 \cap b_{20}) = \lambda_{R_0}(b_{10}, b_{20})$.

For simplicity we add the axioms:

APS 8.1 $M = \bigcup_{a \in Q} a.$

APS 8.2 $M = \bigcup_{b \in R} b.$

The following theorem can be proved.

Theorem 2.1 *The function λ_S for S is uniquely determined by λ_Q and by the particular values $\lambda_S(a \cap b_0, a \cap b)$ for $(a, b_0) \in C$ and $b \in R$ with $b \subset b_0$.*

The role of a physical experiment is to "measure" these particular values of the function λ_S. Therefore we define the set

$$F = \{(b_0, b) : b_0 \in R_0, b_0 \neq \phi, b \in R, b \subset b_0\}$$

F is called the set of "effect procedures" or "questions". The real valued function

$$\mu(a, f) = \mu(a, (b_0, b)) = \lambda_S(a \cap b_0, a \cap b)$$

is defined on $C = \{(a, f) : f = (b_0, b) \in F \text{ and } (a, b_0) \in C\}$

The function μ plays a central role in the statistical description of physical systems.

By the development of the theory up to this point the physical interpretation of quantum mechanics is already established. For this interpretation there is no need of such concepts as states, observables, properties etc. On the contrary, the latter notions will be introduced in the sequel as *derived* notions the interpretation of which are given by the derivation of the notions and the physical interpretation of the preparation and registration procedures.

3. Ensembles and effects

To be short we shall not discuss in detail additional axioms for the set C. The following assumption results as a theorem from physically more plausible axioms:

APS 5.3 The relation $\mu(a_1, f) = \mu(a_2, f)$ for all $f \in \mathcal{F}$ for which $\mu(a_1, f)$ and $\mu(a_2, f)$ are defined is an equivalence relation $a_1 \sim a_2$ on Q.

Let $\mathcal{K}$ be the set of equivalence classes of Q' ($Q' = Q \setminus \{\phi\}$). The elements of $\mathcal{K}$ are called *ensembles* or *states.*

The canonical mapping of Q' onto $\mathcal{K}$ will be denoted by φ.

APS 5.1.4 For all $b_0 \in R_0'$ and $w \in \mathcal{K}$ there exists an $a \in w$ such that $(a, b_0) \in C$.

By $\hat{\mu}(w, f) = \mu(a, f)$ for $a \in w \in \mathcal{K}$ a function $\hat{\mu}$ is defined on the whole set $\mathcal{K} \times \mathcal{F}$ according to APS 5.1.4.

It follows that the relation

$$\hat{\mu}(w, f_1) = \hat{\mu}(w, f_2) \text{ for all } w \in \mathcal{K}$$

defines an equivalence relation $f_1 \sim f_2$ on $\mathcal{F}$.

The set of equivalence classes of $\mathcal{F}$ will be denoted by $\mathcal{L}$. The elements of $\mathcal{L}$ are called "effects". Let ψ be the canonical mapping of $\mathcal{F}$ onto $\mathcal{L}$. It is easy to prove

Theorem 3.1 *By* $\tilde{\mu}(w, g) = \hat{\mu}(w, f)$ *for* $f \in g \in \mathcal{L}$

a function $\tilde{\mu} : \mathcal{K} \times \mathcal{L} \to \mathbb{R}$ *is defined with the following properties:*

i) $0 \leqslant \tilde{\mu}(w, g) \leqslant 1$

ii) If $\tilde{\mu}(w_1, g) = \tilde{\mu}(w_2, g)$ *for all* $g \in \mathcal{L}$ *then* $w_1 = w_2$.

iii) If $\tilde{\mu}(w, g_1) = \tilde{\mu}(w, g_2)$ *for all* $w \in \mathcal{K}$ *then* $g_1 = g_2$.

iv) *There exists a* $g_0 \in \mathcal{L}$ *such that* $\tilde{\mu}(w, g_0) = 0$ *for all* $w \in \mathcal{K}$.

v) *There exists a* $g_1 \in \mathcal{L}$ *such that* $\tilde{\mu}(w, g_1) = 1$ *for all* $w \in \mathcal{K}$.

vi) *To each* $g \in \mathcal{L}$ *there exists a* $g' \in \mathcal{L}$ *such that* $\tilde{\mu}(w, g) + \tilde{\mu}(w, g') = 1$ *for all* $w \in \mathcal{K}$.

If it is clear what function we have in mind, we will write μ instead of $\tilde{\mu}$.

We shall not give a detailed description of the physical possibility of mixing preparation and registration procedures and add only the result of corresponding axioms (see [3] III, §2):

For all rational numbers λ with $0 < \lambda < 1$ and all pairs of ensembles $w_1, w_2 \in \mathcal{K}$ there exists an ensemble $w \in \mathcal{K}$ such that

$$\mu(w, g) = \lambda\mu(w_1, g) + (1 - \lambda)\,\mu(w_2, g) \text{ for all } g \in \mathcal{L}.$$

For all rational numbers λ with $0 < \lambda < 1$ and all pairs of effects g_1, $g_2 \in \mathcal{L}$ there exists an effect $g \in \mathcal{L}$ such that

$$\mu(w, g) = \lambda\mu(w, g_1) + (1 - \lambda)\,\mu(w, g_2) \text{ for all } w \in \mathcal{K}.$$

4. Injections of the sets of ensembles and effects into topological vector spaces

For the next steps of the development we shall disregard the special structures concerning preparation and registration procedures and shall be concerned only with the following structure: A mapping $K \times L \xrightarrow{\mu} [0, 1]$, for which the relations of Theorem 3.1 are valid. This is sufficient to embed K and L into vector spaces.

D 1.1 Let $\underline{D}$ be the set of all functions $y : K \to \mathbb{R}$ of the form $y(w) = \sum_{i=1}^{n} \alpha_i \mu(w, g_i)$ with $\alpha_i \in \mathbb{R}, g_i \in L$. $\underline{B}$ may be the set of all functions $x : L \to \mathbb{R}$ of the form $x(g) = \sum_{i=1}^{n} \beta_i \mu(w_i, g)$ with $\beta_i \in \mathbb{R}, w_i \in K$.

$w \to \mu(w, g)$ is an injective mapping $K \to \underline{B}$ and $g \to \mu(w, g)$ an injective mapping $L \to \underline{D}$. We identify K and L with their ranges in $\underline{B}$ and $\underline{D}$ respectively.

Theorem 4.1 *There is a unique real bilinear form on $\underline{B} \times \underline{D}$ which coincides with μ on the subset $K \times L$ of $\underline{B} \times \underline{D}$.*

Because of Theorem 4.1 this bilinear form will be denoted by $\mu(x, y)$ for $(x, y) \in \underline{B} \times \underline{D}$.

Theorem 3.1 iv) implies, that the 0-element of $\underline{D}$ is an element of L: $0 \in L$. Similarly Theorem 3.1 v) implies the existence of an element $\mathbf{1} \in L$ defined by $\mu(w, \mathbf{1}) = 1$ for all $w \in K$. By Theorem 3.1 vi) $g \in L$ implies $\mathbf{1} - g \in L$.

The elements of K and L were defined as equivalence classes (see § 3); therefore the physical inaccuracy (as defined in [1] § 6) for the elements of K is described by the initial topology defined by the mapping $\mu(\ldots, g) : K \to [0, 1]$ for all $g \in L$ and for the elements of L by the initial topology defined by the mappings $\mu(w, \ldots) : L \to [0, 1]$ for all $w \in K$. These topologies are identical with the $\sigma(\underline{B}, \underline{D})$, resp. $\sigma(\underline{D}, \underline{B})$ topologies on the vector spaces $\underline{B}$ resp., $\underline{D}$ induced on the subsets $K \subset \underline{B}$, resp., $L \subset \underline{D}$.

Let $\hat{K}$ denote the completion of K with respect to the uniform structure generated by $\sigma(\underline{B}, \underline{D})$ and $\hat{L}$ the corresponding completion of L.

Let $\bar{K}$ denote the greatest subset of $\hat{K}$, for which $\sigma(\underline{D}, K) = \sigma(\underline{D}, \underline{B})$ and $\sigma(D, \bar{K})$ coincide *on* L. Similarly we denote by $\bar{L}$ the greatest subset of $\hat{L}$ for which $\sigma(\underline{B}, \bar{L})$ and $\sigma(\underline{B}, L) = \sigma(\underline{B}, \underline{D})$ coincide *on* K. It follows $K \subset \bar{K}$ and $L \subset \bar{L}$. There are no physical possibilities to distinguish between the elements of K and $\bar{K}$ and $\bar{L}$ and L, respectively. The structure of a physically meaningful theory should depend only on the sets $\bar{K}$ and $\bar{L}$.

If $\underline{B}^*$ denotes the set of all linear functionals on $\underline{B}$, $\underline{D}$ is a subspace of $\underline{B}^*$ and $\hat{L}$ is the $\sigma(\underline{B}^*, \underline{B})$ – closure of L in $\underline{B}^*$. Let $\tilde{\underline{D}}$ be the set of all $y \in \underline{B}^*$ which are $\sigma(\underline{B}, \underline{D})$-continuous on K. Evidently $\tilde{D}$ is a subspace of $\underline{B}^*$ and $\bar{L} = \hat{L} \cap \tilde{D}$.

$\|y\| = \sup_{w \in K} |\mu(w, y)|$ is a norm on $\tilde{D}$ and $\tilde{D}$ endowed with this norm is a Banach space.

Likewise the set R of all $y \in \underline{B}^*$ bounded on K is a Banach space under the same norm $\|y\| = \sup_{w \in K} |\mu(w, y)|$. For the norm closure D of $\underline{D}$ in R holds $\tilde{D} \subset D \subset R$ and therefore also $\hat{L} \cap D \subset \bar{L}$. $\bar{L} = \hat{L} \cap D$ follows by theorem 4.3 below. $\bar{L}$ denotes the normclosure of L.

With R' as the dual Banach space of R we get $\underline{B} \subset R'$. If $\tilde{R}$ is a subspace of R with $\underline{D} \subset \tilde{R} \subset R$.

$$\|x\|_{\tilde{R}} = \sup_{\substack{y \in \tilde{R} \\ y \neq 0}} \frac{|\mu(x, y)|}{\|y\|}$$

is a norm on $\underline{B}$. Let $B_{\tilde{R}}$ be the completion of $\underline{B}$ with respect to the norm $\|x\|_{\tilde{R}}$. The dual Banach space $B'_{\tilde{R}}$ of $B_{\tilde{R}}$ can be identified with a subspace of R: $\underline{D} \subset \tilde{R} \subset B'_{\tilde{R}} \subset R$.

The norm of $B'_{\tilde{R}}$ in general, coincides with the norm of R only on the subspace $\tilde{R}$. A subspace $\tilde{R}$ with $\underline{D} \subset \tilde{R} \subset R$ will be called symmetric if $\tilde{R} = B'_{\tilde{R}}$. R itself is symmetric. The intersection of symmetric subspaces is symmetric. In addition we proved in [6], that a $\sigma(B'_{\tilde{R}}, B_{\tilde{R}})$-closed subspace T of a symmetric space $\tilde{R}$ is also $\sigma(R, B_R)$-closed and that every $\sigma(R, B_R)$-closed subspace $\tilde{R}$ of R with $\underline{D} \subset \tilde{R}$ is symmetric.

Theorem 4.2 *There exists a symmetric subspace $\tilde{R}$ with $\underline{D} \subset \tilde{R} \subset R$, such that $\sigma(B_{\tilde{R}}, L)$ separates,* i.e. *$x \in B_{\tilde{R}}$ and $\mu(x, g) = 0$ for all $g \in L$ imply $x = 0$. This condition determines $\tilde{R}$ uniquely as the intersection of all symmetric subspaces of R. This uniquely determined subspace $\tilde{R}$ is the $\sigma(R, B_R)$-closure of $\underline{D}$ in R. We have $D \subset \tilde{R}$ and $\|w\|_{\tilde{R}} = 1$ for all $w \in K$ and $\|g\| \leqslant 1$ for all $g \in L$. Since $\tilde{R}$ is the dual Banach space of $B_{\tilde{R}}$, we find*

$$\|y\| = \sup \{|\mu(x, y)| : x \in B_{\tilde{R}} \text{ and } \|x\| = 1\}.$$

Proof: The intersection Δ of all symmetric subspaces of R is the smallest symmetric subspace of R. Since Δ is also the smallest $\sigma(R, B_R)$-closed subspace of R containing $\underline{D}$ we find: Δ is the $\sigma(R, B_R)$-closure of $\underline{D}$ in R.

If $\sigma(B_\Delta, L)$ does not separate the linear span of L is not $\sigma(\Delta, B_\Delta)$-dense in Δ. Thus there is a $\sigma(\Delta, B_\Delta)$-closed subspace $T \neq \Delta$ which is again a symmetric subspace of R in contradiction to Δ being the smallest one. In order to show the uniqueness property consider a symmetric subspace $\tilde{R} \supsetneq \Delta$. Δ is $\sigma(R, B_R)$-closed and since $\sigma(\tilde{R}, B_{\tilde{R}})$ is finer than $\sigma(R, B_R)$, Δ is also $\sigma(\tilde{R}, B_{\tilde{R}})$-closed. Hence there is an $x \in B_{\tilde{R}}$ orthogonal to Δ in contradiction to the assumption that $\sigma(B_{\tilde{R}}, L)$ separates. ∎

The uniquely determined Banach space $B_{\tilde{R}}$ with $\tilde{R} = \Delta$ according to Theorem 4.2 will be denoted by B so that $\tilde{R}$ is equal to B'. K is a subset of B and L a subset of B'. D is the norm closure of $\underline{D}$ in B'.

Let K be the norm closure of $\mathcal{K}$ in B and let L be the $\sigma(B', B)$-closure of $\mathcal{L}$ in B'. Since L is $\sigma(B', B)$-compact, $\sigma(B', B)$ and $\sigma(B', \underline{B})$ coincide on $L \supset \mathcal{L}$ and since L is $\sigma(B', B)$-complete, we have $L = \hat{\mathcal{L}}$.

The elements of K will be also called ensembles or states and the elements of L also effects.

If one introduces mathematical structures describing the physical possibility to mix preparation procedures and, likewise to mix registration procedures (see the end of § 3), the sets K and L can be proved to be convex. In the sequel convexity of K and L will be assumed.

It is easy to prove : $\|w\| = 1$ and $\mu(\mu, \mathbf{1}) = 1$ for all $w \in K$ and

$$\|y\| = \sup_{w \in \mathcal{K}} |\mu(w, y)| = \sup_{w \in K} |\mu(w, y)|$$

for all $y \in B'$; $\|g\| \leqslant 1$ for all $g \in L$; $0 \leqslant \mu(w, g) \leqslant 1$.
for all $w \in K$, $g \in L$

With D' as the dual Banach space of D it is possible to identify B with a subspace of D'. Hence $\mathcal{K} \subset K \subset D'$.

If $\bar{A}^{\sigma}$ denotes the $\sigma(D', D)$-closure of a subset A of D' we have. $K \subset \bar{\mathcal{K}}^{\sigma}$ and $\bar{K}^{\sigma} = \bar{\mathcal{K}}^{\sigma}$. $\bar{\mathcal{K}}^{\sigma}$ is also convex. It it not difficult to prove $\hat{\mathcal{K}} = \bar{\mathcal{K}}^{\sigma} = \bar{K}^{\sigma}$.

According to the general argumentation in [1] § 9 we assume Q, R, R_0 and thus also $\mathcal{K}$ and $\mathcal{L}$ to be denumberable sets. Hence B and D are separable Banach spaces and the topologies $\sigma(B', B)$ and $\sigma(D', D)$ are metrizable on norm bounded subsets of B' and D', respectively.

Theorem 4.3: *$\tilde{D} = D$ and therefore $\bar{\mathcal{L}} = \hat{\mathcal{L}} \cap D = L \cap D$.*

Proof: An element $l \in \tilde{D}$ is a linear form on $\underline{B}$ which is $\sigma(\underline{B}, \mathcal{L})$-continuous on $\mathcal{K}$. $\mathcal{K}$ is a subset of the $\sigma(D', D)$-compact set $\bar{K}^{\sigma}$. Therefore the topologies $\sigma(D', \mathcal{L})$ and $\sigma(D', D)$ coincide on $\mathcal{K}$. Thus l has a unique continuous extension onto $\bar{K}^{\sigma}$. Since D' is the linear span of $\bar{K}^{\sigma}$, l can be linearly extended onto D'.

We will prove that l is $\sigma(D', D)$-continuous on D'. l is $\sigma(D', D)$-continuous iff the subspace F of all $x \in D'$ with $l(x) = 0$ is $\sigma(D', D)$-closed. F is $\sigma(D', D)$-closed iff $F \cap D'_{(1)}$ ($D'_{(1)}$ being the unit ball of D') is $\sigma(D', D)$-closed. Let $x \in D'_{(1)}$ be a cluster point of $F \cap D'_{(1)}$.

Since $\sigma(D', D)$ is metrizable on $D'_{(1)}$ there exists a sequence $x_\nu \in F \cap D'_{(1)}$ with $x_\nu \to x$ in the $\sigma(D', D)$-topology. Since every $x_\nu \in D'$ has a decomposition $x = \alpha w - \beta u$ with $w, u \in \bar{K}^{\sigma}$, $\alpha, \beta \geqslant 0$ and $\|x\| = \alpha + \beta$, we get $x_\nu = \alpha_\nu w_\nu - \beta_\nu u_\nu$ with $w_\nu, u_\nu \in \bar{\mathcal{K}}^{\sigma}$ $\alpha_\nu + \beta_\nu \leqslant 1$. There exists a subsequence (which for simplicity is also denoted by x_ν) with $\alpha_\nu \to \alpha$, $\beta_\nu \to \beta$. With $\bar{x}_\nu = \alpha w_\nu - \beta u_\nu$ we get $\|x_\nu - \bar{x}_\nu\| \to 0$ and therefore also $\bar{x}_\nu \to x$ in the $\sigma(D', D)$-topology. Since l is norm continuous from $l(\bar{x}_\nu) - l(x_\nu) = l(\bar{x}_\nu)$ follows $l(\bar{x}_\nu) \to 0$.

Since $\bar{K}^{\sigma}$ is $\sigma(D', D)$-compact, there exists a subsequence of the sequence $\bar{x}_\nu$

(which again is denoted by $\bar{x}_\nu$), so that $w_\nu \to w$ and $u_\nu \to u$ in the $\sigma(D', D)$-topology. Hence $\bar{x}_\nu \to x = \alpha w - \beta u$. Since l is $\sigma(D', D)$-continuous on $\bar{K}^\sigma$ we get $l(w_\nu) \to l(w)$ and $l(u_\nu) \to l(u)$ and therefore $l(\bar{x}_\nu) = \alpha l(w_\nu) - \beta l(u_\nu) \to \alpha l(w) - \beta l(u) = l(x)$. According to $l(\bar{x}_\nu) \to 0$ we get $l(x) = 0$, i.e. $x \in F$.

Since l is $\sigma(D', D)$-continuous on D', we have $l \in D$, i.e. $\tilde{D} \subset D$. We have already noted above $D \subset \tilde{D}$. ■

Similarly, the following theorem can be proved (L° denotes the norm-interior of L:):

Theorem 4.4 *The two relations $K = B \cap \bar{K}^\sigma$ and $L^\circ \neq \phi$ imply $\breve{K} = K$.*

Axiom AV 1.2s (introduced in § 5) implies $L^\circ \neq \phi$ and using Theorem 5.10 and Theorem 5.9 also $K = B \cap \bar{K}^\sigma$.

We will report on some well-known mathematical results without giving the proofs (for proofs see e.g. [6] and [7]).

$B_+ = \bigcup_{\lambda \geqslant 0} (\lambda K)$ is a cone by which B becomes an ordered vector space.

B_+ is norm- and $\sigma(B, B')$-closed. K is a base of B_+ so that B is a base norm space with K as base, i.e. $\overline{\text{co}}\,(K \cup (-K))$ is the unit ball of B. The cone $\{y : y \in B'$ and $\mu(x, y) \geqslant 0$ for all $x \in B_+\}$ is called the dual cone B'_+ to B_+. B'_+ is $\sigma(B', B)$-closed. $B_+ = \{x \mid x \in B$ and $\mu(x, y) \geqslant 0$ for all $y \in B'_+\}$. B' is an order unit space, i.e. the order intervall $[-1, 1]$ is the unit ball of B'. The set L is a subset of the order intervall $[0, 1]$. It follows $B = B_+ - B_+$ and $B' = B'_+ - B'_+$. Every positive linear form on B is an element of B'_+, i.e. is norm-continuous. B'_+ is $\sigma(B', B)$-complete.

For $x \in B$ we have $\|x\| = \inf\{\alpha + \beta : \alpha, \beta \geqslant 0, x = \alpha w_1 - \beta w_2$ and $w_1, w_2 \in K\}$. Every $x \in B$ has a decomposition $x = \alpha w_1 - \beta w_2$ with $\alpha, \beta \geqslant 0$ and $w_1, w_2 \in K$ so that $\alpha + \beta \leqslant \|x\| \leqslant \alpha + \beta + \epsilon$ for arbitrary $\epsilon > 0$. For every $\epsilon > 0$ there exists also a decomposition $x = \alpha w_1 - \beta w_2$, so that there is an $y_0 \in [-1, 1]$, with $\mu(w_2, y_0) = -1$ and $\mu(w_1, y_0) \geqslant 1 - \epsilon$ and an $y'_0 \in [-1, 1]$ with $\mu(w_1, y'_0) = 1$ and $\mu(w_2, y'_0) \leqslant -1 + \epsilon$.

We will give some general relations for the faces of K. The exposed faces of K are of the form

$$K_0(y) = \{w : w \in K \text{ and } \mu(w, y) = 0\} \text{ for a } y \in B'_+.$$

Let $J = \{y : y \in [0, 1]$ and $K_0(y) \neq \phi\}$, then

$$K = \{x : \mu(x, 1) = 1 \text{ and } \mu(x, y) \geqslant 0 \text{ for all } y \in J\}.$$

All sets $K_0(N) = \{w : w \in K$ and $\mu(w, y) = 0$ for $y \in N \subset B'_+\}$ are exposed faces (B' is $\sigma(B', B)$-separable!).

An exposed face F_1 of K is called orthogonal to another exposed face F_2, $F_1 \perp F_2$, iff there exists a $y \in [0, 1]$ with $F_1 \subset K_0(y)$ and $F_2 \subset K_0(1 - y)$.

We introduce the following sets of faces:

$W = \{F : F \text{ is a closed face of } K\}$,

$W_{B'} = \{K_0(N) : N \subset B'_+\}$,

$W_L = \{K_0(N) : N \subset L\}$,

It follows immediately $W \supset W_{B'} \supset W_L$ and $W_{B'} = \{K_0(N) \mid N \subset [0, 1]\}$. $W_{B'}$ equals also the set of all exposed faces of K.

We define

$L_0(k) = \{y : y \in L \text{ and } \mu(w, y) = 0 \text{ for } w \in k \subset K\}$,

$\tilde{L}_0(k) = \{y : y \in [0, 1] \text{ and } \mu(w, y) = 0 \text{ for } w \in k \subset K\}$,

$U = \{L_0(k) : k \subset K\}$,

$V = \{\tilde{L}_0(k) : k \subset K\}$.

The sets W, $W_{B'}$, W_L, U, V are complete lattices of sets, the set theoretical and lattice theoretical intersections coincide.

$K_0, L_0, \tilde{L}_0$ can be interpreted as mappings $K_0 : V \to W_{B'}$, $K_0 : U \to W_L$, $\tilde{L}_0 : W_{B'} \to V$, $L_0 : W_L \to U$. These mappings are dual-isomorphisms of the lattices.

Let $C(w)$ denote the closed face of K generated by $w \in K$. Every closed face of K is of the form $C(w)$ for a suitable w.

As a conclusion of this section we may formulate: The structure of ensembles and effects for physical systems can be described by a general base norm space B; the set of ensembles can be identified with the base K of B. The set of effects can be identified with a subset L of B' such that $L \subset [0, 1]$ and the $\sigma(B', B)$-closed subspace generated by L is B'.

5. Laws of preparation and registration

The structure of the ensembles and effects gained up to now is much to general as to characterize microsystems. It is necessary to add general laws for the sets K and L. As we have seen these sets are "physically" determined so that every law for K and L is on principle a law concerning the possibilities of preparation and registration. A more detailed discussion of this question can be found in [6], [7] and [8].

This section contains certain improvements and a summarizing discussion of the mathematical aspect of the mentioned laws of preparation and registration. For the epistemological aspect we refer to [2], [6] and [8].

We will not prove all theorems of this sections since most of them were already proved in [6], [8].

Consider the following relation

AV 1.1 For all pairs $g_1, g_2 \in L$ there is an element $g \in L$ with $g \geqslant g_1$, $g \geqslant g_2$ and $K_0(g_1) \cap K_0(g_2) \subset K_0(g)$.

It is not difficult to prove:

Theorem 5.1 *The following relations are equivalent:*

i) *AV 1.1*

ii) *The elements of U, i.e. the sets $L_0(k) = \{g : g \in L$ and $\mu(w, g) = 0$ for all $w \in k \subset K\}$ are upwards directed subsets of B'.*

iii) *The sets*

$$L_1(k) = \{g : g \in L \text{ and } \mu(w, g) = 1 \text{ for all } w \in k \subset K\}$$

are downwards directed subsets of B'.

iv) *The elements of U have a greatest element.*

v) *The sets $L_1(k)$ have a smallest element.*

We introduce AV 1.1 as a first law, i.e. as an axiom of the theory. Then ii) to v) are theorems of the theory.

The greatest element of $L_0(k)$ is denoted by $eL_0(k)$ and is called *decision effect.* The set of all decision effects will be denoted by G.

The following relations can be proved:

$$L_0(k) = \{g : g \in L \text{ and } g \leqslant e\, L_0(k)\}, \tag{5.1}$$

$$K_0 L_0(k) = K_0(e\, L_0(k)), \tag{5.2}$$

$$L_0(k) = L_0 K_0(e\, L_0(k)), \tag{5.3}$$

for all $e \in G$: $\{g \mid g \in L \text{ and } g \leqslant e\} = L_0 K_0(e)$. (5.4)

The structures mentioned at the end of § 4 and (5.1) allow to prove:

Theorem 5.2 *G is a complete lattice. The mapping $L_0(k) \to e\, L_0(k)$ is an isomorphism $e : U \to G$. The mapping $e\, L_0(k) \to K_0(e\, L_0(k))$ is a dual-isomorphism $G \to W_L$.*

Theorem 5.3 *For $g \in L$ and $e \in G$ the following two relations are equivalent:*

i) $g \leqslant e$,

ii) $K_0(g) \supset K_0(e)$.

To each subset $l \subset L$ there exists exactly one $e \in G$ such that $K_0(l) = K_0(e)$, i.e. a smallest element $e \in G$ with $e \geqslant g$ for all $g \in l$.

With $\partial_e L$ denoting the set of extremal points of L follows

Theorem 5.4 $G \subset \partial_e L$.

Other theorems are:

Theorem 5.5 *If $e_1, e_2 \in G$ then*

$$K_0(e_1) \cap K_0(e_2) = K_0(e_1 \vee e_2).$$

Theorem 5.6 *For $e_1, e_2 \in G, g \in L$ we have:*

$$g \leqslant e_1, g \leqslant e_2 \Rightarrow g \leqslant e_1 \wedge e_2.$$

Theorem 5.7 *For a subset $A \subset G$ the infimum $\bigwedge_{e \in A} e$ of A in G is also the infimum of A in L. If A is directed downwards the $\sigma(B', B)$-limes of A is $\bigwedge_{e \in A} e$.*

We will now discuss the following relations:

AV 1.2 For every $g_0 \in L$ and $w_0 \in K$ with $\mu(w_0, g_0) = 0$ and every $y \in [0, 1]$ with $\mu(w_0, y) =$ exists an element $g \in L$ with $\mu(w_0, g) = 0$ and $\mu(w, g) \geqslant 1 - \delta(\epsilon)$ for all $w \in K$ with $\mu(\tilde{w}, y) \leqslant 1 - \epsilon$; the function $\delta(\epsilon)$ can be chosen such that $\delta(\epsilon) \to 0$ if $\epsilon \to 0$.

AV 1.2s $L = [0, 1]$.

AV 2: If F is an exposed face of K, $w_0 \in F$, $\tilde{w} \in K$ and $\tilde{w} \notin F$ then there exists a sequence $g_\nu \in L$ with $\mu(w_0, g_\nu) \to 0$ and $\mu(w, g_\nu) > \delta > 0$ for all g_ν.

AV 2s: If F is a closed face of K, $w_0 \in F$, $\tilde{w} \in K$ and $\tilde{w} \notin F$ then there exists a sequence $g_\nu \in L$ with $\mu(w_0, g_\nu) \to 0$ and $\mu(\tilde{w}, g_\nu) > \delta > 0$ for all g_ν.

AV 2f: Every finite dimensional face of K is an exposed face of K.

For the proofs of theorems 5.8 to 5.10 the assumption AV 1.1 is not necessary.

Theorem 5.8 *The following relations are equivalent:*

i) *If F is an exposed face of K, $w \in K$, $w \notin F$ then there exists a $g \in L$ with $K_0(g) \supset F$ and $\mu(w, g) \neq 0$.*

ii) *The sets W_L and $W_{B'}$ (as defined in § 4) coincide.*

iii) *Let F_1 and F_2 be exposed faces of K with $F_1 \subsetneqq F_2$. $F_1 = C(w_1)$, $F_2 = C(w_2)$ imply the existence of a $g \in L$ with $\mu(w_1, g) = 0$ and $\mu(w_2, g) \neq 0$.*

iv) *Let F_1 and F_2 be exposed faces of K with $F_1 \subsetneqq F_2$. $F_1 = C(w_1), F_2 = C(w_2)$ imply $L_0(w_1) \neq L_0(w_2)$.*

v) *Let F_1 and F_2 be exposed faces of K, $F_1 = C(w_1), F_2 = C(w_2)$. The relation $L_0(w_1) = L_0(w_2)$ implies $F_1 = F_2$.*

vi) *For every exposed face F of K there exists a $g \in L$ with $F = K_0(g)$.*

vii) *AV 2.*

Theorem 5.9 *AV 2 implies that the $\sigma(B', B)$-closed cone generated by L is B'_+.*

Theorem 5.10 [*AV 1.2 and AV 2*] $\Leftrightarrow$ *AV 1.2s.*

If we assume AV 1.2s (but not AV 1.1) we find: $D \cap [0, 1]$ is $\sigma(B', B)$-dense in $[0, 1]$. B is a Banach subspace of D'. $\bar{K} = K$ and $\bar{L} = D \cap [0, 1]$.

Now, assuming AV 1.1 *and* AV 1.2s the following theorem can be proved with the help of the general structures mentioned in the second part of § 4.

Theorem 5.11 $\overline{\text{co}}^\sigma G = [0, 1] = L$, $\partial_e L \subset \bar{G}^\sigma$.

For $e \in G$ we have $L_0 K_0(e) = [0, e]$. ($-^\sigma$ denotes the $\sigma(B', B)$-closure).

The stronger relation $G = \partial_e[0, 1] = \partial_e L$ could not be proved without additional assumptions.

It is not difficult to prove $\sup_{w \in K} \mu(w, e) = 1$ for all $e \in G$ with $e \neq 0$. However, we did not succeed in proving that there exists a $w \in K$ with $\mu(w, e) = 1$. Because of the physical inaccuracy it is impossible to distinguish physically the case $\sup_{w \in K} \mu(w, e) = 1$ from the case, that there exists a $w \in K$ with $\mu(w, e) = 1$. Therefore the following relation is only a mathematical idealization and has no physical significance.

AVid To each $e \in G$ there exists a $\tilde{w} \in K$ such that $\mu(\tilde{w}, e) = \sup_{w \in K} \mu(w, e)$.

Again assuming AV 1.1 and AV 1.2s we can prove:

Theorem 5.12 *The following relations are equivalent:*

i) AVid,
ii) $1 - G \subset G$,
iii) $1 - G = G$,
iv) $e \in G, e \neq 0 \Rightarrow K_0(1 - e) \neq \phi$,
v) $e_1, e_2 \in G$ *and* $e_1 \leqslant e_2$ *imply* $e_2 - e_1 \in G$.

Under the assumption of AV 1.1, AV 1.2s, AVid we get the following theorems

Theorem 5.13 *All elements of G are exposed points of* $L = [0, 1]$.

Theorem 5.14 *The mapping* $e \to e^\perp = 1 - e$ *defines an orthocomplementation* $\perp$ *in the lattice G.*

We define e_1 orthogonal $e_2 (e_1 \perp e_2)$ iff $e_1 \leqslant e_2^\perp$. Since $g' = 1 - g \in L$ we have $\{K_0(g) : g \in L\} = \{K_1(g) : g \in L\}$ with $K_1(g) = \{w : w \in K, \mu(w, g) = 1\}$. Hence $K_1(e) = K_0(e^\perp)$ and $K_1 : G \to W_2$ defines an isomorphism by Theorem 5.2

Two exposed faces F_1, F_2 of K are orthogonal in the sense of § 4, iff (with $F_1 = K_1(e_1), F_2 = K_1(e_2)$) $e_1 \perp e_2$.

Theorem 5.15 $e_1, e_2 \in G$ *and* $e_1 + e_2 \leqslant 1$ *imply* $e_1 \perp e_2$. $e_1 \perp e_2$ *implies* $e_1 \vee e_2 = e_1 + e_2$.

Theorem 5.16 *The supremum* $\bigvee_{e \in A} e$ *of a subset* $A \subset G$ *is also the supremum of A in* $L = [0, 1]$.

Theorem 5.17 $A \subset G$ *and* $\sum_{e \in A} e \leqslant 1$ *implies, that the elements of A are*

pairwise orthogonal and that A is denumerable. Moreover $\sum_{e\in A} e = \bigvee_{e\in A} e$

Theorem 5.18 *In every subset* $A \subset G$ *there exists a denumerable subset* $\{e_\nu\}$ *with* $\bigvee_{e\in A} e = \bigvee_\nu e_\nu$. *An equivalent relation is valid for* $\wedge$.

Theorem 5.19 *The operations* $\wedge, \vee, \perp$ *are continuous in the* $\sigma(B', B)$-*-topology; that means more precisely:*

$e_\nu \to e$ *implies* $e_\nu^\perp \to e^\perp$.

$e_\nu \to e, e'_\mu \to e'$ *and* $\inf_{w\in K} \{\mu(w, (e_\nu \vee e'_\mu)^\perp + \frac{1}{2}(e_\nu + e'_\mu))\} \geqslant \beta > 0$

imply $e_\nu \vee e'_\mu \to e \vee e'$.

$e_\nu \to e, e'_\mu \to e'$ *and* $\inf_{w\in K} \{\mu(w, e_\nu \wedge e'_\mu + \frac{1}{2}(e_\nu^\perp + e'^\perp_\mu))\} \geqslant \gamma > 0$

imply $e_\nu \wedge e'_\mu \to e \wedge e'$.

The set $\mathcal{R}(b_0) = \{b : b \in \mathcal{R}_0, b \subset b_0 \in \mathcal{R}_0\}$ is a Boolean algebra of sets (compate § 2). With $\psi : \mathcal{F} \to \mathcal{L}$ as defined in § 3 we define a mapping ψ_{b_0} : $\mathcal{R}(b_0) \to \mathcal{L}$ by $\psi_{b_0}(b) = \psi(b_0, b)$. This mapping is an additive measure on the Boolean algebra $\mathcal{R}(b_0)$. $\mathcal{R}(b_0)$ represents physically a registration method with all its possible reponses b.

As an idealization of the mapping $\psi_{b_0} : \mathcal{R}(b_0) \to \mathcal{L}$ we define the concept of *coexistent* effects and the concept of an *observable* (see [4], [6] and [3]):

A subset $A \subset L$ is called a set of coexistent effects iff there exists a Boolean algebra Σ with an additive measure $F : \Sigma \to L$ so that $A \subset F\,\Sigma$.

Correspondingly a subset $A \subset G$ is called a set of commensurable decision effects iff there exists a Boolean algebra Σ with an additive measure $F : \Sigma \to G$ so that $A \subset F\,\Sigma$.

An observable is a complete Boolean algebra with a σ-additive measure $F : \Sigma \to L$. A decision observable is a complete Boolean algebra with a σ-additive measure $F : \Sigma \to G$.

Theorem 5.20 *The following relations for* $g_1, g_2 \in L$ *are equivalent:*

i) $\{g_1, g_2\}$ *coexistent.*

ii) *there exist elements* $g'_1, g'_2, g_{12} \in L$ *such that* $g_1 = g'_1 + g_{12}$, $g_2 = g'_2 + g_{12}, g'_1 + g'_2 + g_{12} \in L$.

Theorem 5.21 *The following relations for* $e \in G$ *and* $g \in L$ *are equivalent:*

i) $\{g, e\}$ *coexistent.*

ii) *there exist* $g_1, g_2 \in L$ *such that* $g = g_1 + g_2$ *and* $g_1 \leqslant e, g_2 \leqslant e^\perp$.

iii) *there exist* $g'_1, g'_3 \in L$ *such that* $e = g'_1 + g'_3$ *and* $g'_1 \leqslant g, g'_3 \leqslant 1 - g$.

Theorem 5.22 *Let $F : \Sigma \to G$ be an additive and effective measure on the Boolean algebra Σ. Then $F\Sigma$ is a Boolean subalgebra of G and $F : \Sigma \to F\Sigma$ an isomorphism.*

Theorem 5.23 *The following conditions for decision effects e_1, e_2 are equivalent:*

i) $\{e_1, e_2\}$ *coexistent.*

ii) $\{e_1, e_2\}$ *commensurable.*

iii) *The orthocomplemented sublattice Γ of G generated by $\{e_1, e_2\}$ is Boolean*

iv) $e_1 = (e_1 \wedge e_2) \vee (e_1 \wedge e_2^{\perp})$

Theorem 5.24 *Let A_1 and A_2 be subsets of G such that e_1 and e_2 are commensurable for all $e_1 \in A_1$ and $e_2 \in A_2$. Then also $\bigwedge_{e \in A_2} e$ and $\bigvee_{e \in A_2} e$ are commensurable with all $e_1 \in A_1$. If e is commensurable with all $e_1 \in A_1$, the $e^{\perp}$ is commensurable with all $e_1 \in A_1$.*

Theorem 5.25 *The following relations are equivalent for a subset $A \subset G$:*

i) *A is coexistent*

ii) *A is commensurable*

iii) *The elements of A are pairwise coexistent.*

iv) *The elements of A are pairwise commensurable.*

v) *The orthocomplemented sublattice generated by A is Boolean.*

vi) *The orthocomplemented and complete sublattice generated by A is Boolean.*

Theorem 5.26 *If $\{e_\nu\}$ is a sequence of commensurable decision effects and $e_\nu \to e \in G$ in the $\sigma(B', B)$-topology then e is commensurable with all $\bar{e} \in G$ which are commensurable with all e_ν. There exists a subsequence e_{ν_k} such that $e = \bigwedge_{n=1}^{\infty} \tilde{e}_n$ with $\tilde{e}_n = \bigvee_{k=n}^{\infty} e_{\nu_k}$. (Especially, as a consequence of these relations every complete Boolean subalgebra of G is $\sigma(B', B)$-closed in G.)*

Next we want to prove the orthomodularity of G. Therefore we introduce the following relations for an orthocomplemented lattice:

The modular relation $M(a, b) : c < b$ implies $(c \vee a) \wedge b = c \vee (a \wedge b)$.

The compatibility relation $C(a, b) : a = (a \wedge b) \vee (a \wedge b^{\perp})$.

The following theorem valid in an orthocomplemented lattice is well-known:

Theorem 5.27 *The following relations are equivalent:*

i) *$a \perp b$ implies $M(a, b)$.*

ii) $M(a, a^{\perp})$,

iii) *$a < b$ implies $C(b, a)$,*
iv) *$a < b$ implies $a = b \wedge (a \vee b^{\perp})$,*
v) *$C(a, b)$ implies $C(b, a)$,*
vi) *$a \perp b, a \perp c$ and $a \vee b = a \vee c$ imply $b = c$,*
vii) *$a \perp b$, $a \perp c$ and $a \vee b = a \vee c$ and $b \leqslant c$ imply $b = c$*
viii) *$b \perp a$ and $a \vee b = 1$ imply $b = a^{\perp}$.*

If one of the relations of Theorem 5.27 is valid (and then also all the others) the lattice is called orthomodular.

Theorem 5.28 *G is orthomodular.*

Proof: $e_1 \perp e, e_2 \perp e$ and $e \vee e_1 = e \vee e_2$ imply $e + e_1 = e + e_2$ according to Theorem 5.15; it follows $e_1 = e_2$, i.e. vi) of Theorem 5.27. ∎

The next law, relation AV 3, is only of interest if there are non-commensurable decision effects. We introduce a "distance" in G:

$$\Delta(e_1, e_2) = \max\{\inf_{w \in K_1(e_1)} \mu(w, 1 - e_2); \inf_{w \in K_1(e_2)} \mu(w, 1 - e_1)\}.$$

$\Delta(e_1, e_2) = 1$ is equivalent to $e_1 \perp e_2$.

For commensurable e_1, e_2 holds $\Delta(e_1, e_2) = 0$ or $\Delta(e_1, e_2) = 1$.

Theorem 5.29 *The following relations for elements $e_i \in G$ are equivalent:*

i) $e_2 \perp e_3, \Delta(e_1, e_2 + e_3) \neq 0 \Rightarrow e_3 \nleqslant e_1 \vee \nu_2$,
ii) $e_2 \perp e_3, \Delta(e_1, e_2 + e_3) \neq 0 \Rightarrow e_3 \wedge (e_1 \vee e_2) = 0$,
iii) $e_2 \leqslant e_4, \Delta(e_1, e_4) \neq 0, e_1 \vee e_2 = e_1 \vee e_4 \Rightarrow e_2 = e_4$,
iv) $e_2 \leqslant e_4, \Delta(e_1, e_4) \neq 0, e_2 \neq e_4 \Rightarrow e_4 \nleqslant e_1 \vee e_2$,
v) $e_3 \nleqslant e_2, \Delta(e_1, e_2 + e_3) \neq 0 \Rightarrow e_3 \nleqslant e_1 \vee \nu_2$,
vi) $e_3 \leqslant e_1 \vee e_2, \Delta(e_1, e_2 \vee e_3) \neq 0 \Rightarrow e_3 \leqslant e_2$,
vii) $e_3 \leqslant e_1 \vee e_2, e_3 \leqslant 1 - e_2, \Delta(e_1, e_2 \vee e_3) \neq 0 \Rightarrow e_3 = 0$.
viii) $e_2 \leqslant e_1, \Delta(e_1 \wedge (e_2 \vee e_3), e_3 - e_1 \wedge e_3) \neq 0 \Rightarrow (e_2 \vee e_3) \wedge e_1 = e_2 \vee (e_3 \wedge e_1)$,
ix) $e_2 \leqslant e_1 \leqslant e_2 \vee e_3, \Delta(e_1, e_3) \neq 0 \Rightarrow e_1 = e_2$.

For commensurable $e_1, e_2 \in G$ $\Delta(e_1, e_2) \neq 0$ implies $\Delta(e_1, e_2) = 1$, i.e. $e_1 \perp e_2$. If all elements of G are commensurable, i.e. G is a Boolean algebra, then iii) of Theorem 5.29 is valid since G is orthomodular. This is a mathematical motivation of the "law" AV 3 which is assumed to hold in the sequel. For epistomological aspects of AV 3 we refer to [6], [7], [8].

AV 3 One of the relations i) to ix) of Theorem 5.29.

With AV 3 follows

Theorem 5.30 $\Delta(e_1, e_3 - e_1 \wedge e_3) \neq 0$ *implies the modular relation* $M(e_3, e_1)$, i.e.

$$(e \vee e_3) \wedge e_1 = e \vee (e_3 \wedge e_1) \quad \textit{for all} \quad e \leqslant e_1.$$

Theorem 5.31 *If $K_1(e_1)$ or $K_1(e_3)$ is $\sigma(B, B')$-compact then $M(e_3, e_1)$ and $M(e_1, e_3)$. If $K_1(e_1)$ or $K_1(e_3)$ is $\sigma(B, B')$-compact and $e_2 \leqslant e_1 \leqslant e_2 \vee e_3$. $e_1 \wedge e_3 = 0$ then $e_1 = e_2$.*

In particular, $K_1(e)$ is $\sigma(B, B')$-compact if $K_1(e)$ is finite dimensional. If e_3 is an atom of the lattice G and $K_1(e_3)$ finite dimensional (e.g. an extremal point of K), we get from Theorem 5.31: $e_2 \leqslant e_1 \leqslant e_2 \vee e_3$ implies $e_1 = e_2$ or $e_1 \wedge e_3 \neq 0$; $e_1 \wedge e_3 \neq 0$ is equivalent to $e_3 \leqslant e_1$ (e_3 is an atom!) and therefore $e_1 = e_1 \vee e_3 \geqslant e_2 \vee e_3$, i.e. $e_1 = e_2 \vee e_3$.

Theorem 5.32 *If $K_1(e)$ is $\sigma(B, B')$-compact then the order intervall $[o, e]$ in G is a complete orthocomplemented and modular lattice.*

The last law to be considered next is called the law of quantization and describes that microsystems have so-called "discrete" states. We introduce the following relations

AV 4 For every exposed face F of K there exists another exposed face F_1 of finite dimension with $F_1 \subset F$.

AV 4s Every exposed face F of K is the supremum of an increasing sequence of finite dimensional exposed faces $F_\nu : F = \bigvee_\nu F_\nu ; F_{\nu+1} \supset F_\nu$.

Theorem 5.33 *AV 4 implies: The lattices G and W_L are atomic. An atom of W_L is a finite dimensional face of K.*

Theorem 5.34 *AV 4 and AV 2f imply: the atoms of W_L are the extremal points of K.*

Theorem 5.35 *AV 4 and AV 3 imply: If p is an atom of G and $e_2 \leqslant e_1 \leqslant e_2 \vee p$ then $e_1 = e_2$ or $e_1 = e_2 \vee p$.*

Theorem 5.36 *AV 4 and AV 3 imply: If $e \in G$, $e \neq 0$, $e \neq 1$ and p an atom there exist two atoms $q_1 \leqslant e$, $q_2 \leqslant e^\perp$ such that $p \leqslant q_1 \vee q_2$.*

We will conclude this section by a general discussion of the introduced laws of preparation and registration.

The law AV 1.2s seems to be best founded as long as we have no physical reason to suppose that there are structures of microsystems which in contradiction to AV 1.2s make it impossible to construct sufficient many registration procedures. If we take AV 1.2s for granted all the other axioms AV 1.1, AV 2f, AVid, AV 3 and AV 4 (AV 4s) can be considered as relations concerning the structure of the convex set K. This follows from $[0, 1]$ being the set of *all* affine functions f on K with $0 \leqslant f(w) \leqslant 1$ according to § 4. Therefore, a condition for the set $[0, 1]$ indirectly is a condition for the convex set K. In this sense the structure of K determines the most general structures of the microsystems as carriers of the physical interaction between preparation and registration apparatuses.

We did not prove that under the assumption of AV 1.2s the axioms AV 1.1, AV 2f, AV 3 and AV 4 (AV 4s) are independent. Of course, AV 4 is independent of all the other axioms because there are models where (all decision effects being commensurable) all axioms with the exception of AV 4 are valid. The independence of the remaining axioms AV 1.1, AV 2f, AVid, AV 3 is an open question. It might well be that one of them can be proved. For instance, as can be easily seen AVid is valid if B is finite dimensional. AVid might even be true in the general case or could be a consequence of AV 1.1 and AV 2s which is a stronger version of AV 2f. Another question is whether AV 1.1, AV 2s and AVid imply AV 3. The well-known examples of orthomodular lattices with AV3 not being fulfilled represent no proof of the independence of AV 3 since in our case the lattice G must be considered as a particular subset of $[0, 1]$.

§ 6. The representation of K and L by operators in Hilbert spaces

It is impossible to summarize here the long way in proving the representation theorem 6.13 stated below. We will only comment some crucial points on this way (for details see [7]). In the proof of a corresponding theorem given in [6] and [8] we made some additional but unnecessary assumptions. Indeed, for the proof of Theorem 6.13 only assumptions AV 1.1, AV 1.2s, AV 2f, AVid, AV 3 and AV 4s are presupposed.

The first step of the proof is a reduction of the spaces B, B' (not only of the lattice G!) by the center of G. The center Z of G is defined as the set of all $e \in G$ which are commensurable with all other elements of G. Z is a complete Boolean algebra.

[6] contained an error concerning the center Z. We will therefore prove the following theorem:

Theorem 6.1 *If G is atomic then also Z is atomic. (Only AV 1.1, AV 1.2s, AVid are presupposed.)*

Proof: Assuming Z not to be atomic there exists a $q \in Z$ such that q comprises no atom of Z. Let $G(q) = \{e \mid e \in G \text{ and } e \leqslant q\}$.

Since G is atomic there exists an atom p of G in $G(q)$. We define the following subsets of Z:

$$Z(q;p) = \{\tilde{q} : \tilde{q} \in Z,\ \tilde{q} \leqslant q,\ \tilde{q} \geqslant p\},$$
$$Z(q) \quad = \{\tilde{q} : \tilde{q} \in Z,\ \tilde{q} \leqslant q\}.$$

$Z(q)$ is a Boolean algebra with q as unit element. $Z(q;p) \subset Z(q)$. The following relations are valid for $Z(q;p)$ $(q_1, q_2 \in Z(q))$:

(α) if $q_1 \in Z(q, p)$, $q_2 \in Z(q, p)$ then $q_1 \wedge q_2 \in Z(q, p)$.

(β) if $q_1 \geqslant q_2, q_2 \in Z(q, p)$ then $q_1 \in Z(q, p)$.

(γ) if $q_1 \notin Z(q, p)$ then $q_1^{\perp} \wedge q \in Z(q, p)$.

Since Z is complete $\bar{q} = \bigwedge_{\tilde{q} \in Z} \tilde{q} \in Z(q)$. It follows $p \leqslant \bar{q}$ and therefore $\bar{q} \in Z(q; p)$. Since according to our assumption there is no atom in $Z(q)$ there exists an element $\tilde{q} \in Z(q)$ such that $0 \neq \tilde{q} \lneqq \bar{q}$. It follows $\tilde{q}^{\perp} \wedge \bar{q} \neq 0$ and $\tilde{q}^{\perp} \wedge \bar{q} \neq \bar{q}$. Because of $\tilde{q} \lneqq \bar{q}$ we find $\bar{q} \in Z(q; p)$ and therefore according to (γ) $\tilde{q}^{\perp} \wedge q \in Z(q; p)$ and according to (α) $\tilde{q}^{\perp} \wedge q \wedge \bar{q} = \tilde{q}^{\perp} \wedge \bar{q} \in Z(q; p)$ in contradiction to $0 \neq \tilde{q}^{\perp} \wedge \bar{q} \neq \bar{q}$ and $\bar{q}$ the smallest element of $Z(q; p)$.■

If Z is atomic the set of atoms of Z is denumerable. Let $\{q_\nu\}$ be the set of the atoms of Z. It follows $\sum_\nu q_\nu = \mathbf{1}$.

The axioms AV 1.1, AV 1.2s, AVid (without AV 4) imply: For all $z \in Z$ the linear span of $K_1(z)$ is a Banachspace $B_z \subset B$. An element $y \in B'$ will be called z-bounded if there is a real number c such that $-cz \leqslant y \leqslant cz$.

The dual Banachspace B_z' of B_z can be identified with the ($\sigma(B', B)$-closed) subspace of all z-bounded elements of B': $B_z' = \{y \mid y \in B' \text{ and } y \text{ } z\text{-bounded}\}$. B_z is a base norm space with the base $K_1(z)$. B_z' is an order unit space with the order unit z. With $K_z = K_1(z)$, $L_z = [o, z]$ the axioms AV 1.1, AV 1.2s and AVid are also valid for B_z, B_z'. If AV 2f, AV 3, AV 4, AV 4s are valid in B, B' then also in B_z, B_z'.

Now, we presuppose also AV 4. Then Z is atomic. For an atom $q_\nu \in Z$ the spaces B_{q_ν}, B'_{q_ν} will de denoted by B_ν, B_ν'. Then $B = \sum_\nu B_\nu$, $B' = \sum_\nu B_\nu'$ in the following sense: The elements $x \in B$ have a unique decomposition $x = \sum_\nu x_\nu, x_\nu \in B_\nu$ and the elements $y \in B'$ have a unique decomposition $y = \sum_\nu y_\nu$, $y_\nu \in B_\nu'$.

For the components w_ν of $w \in K$ holds $w_\nu \in \check{K}_\nu = \bigcup_{0 \leqslant \lambda \leqslant 1} \lambda K_\nu$ and for the components g_ν of $g \in L$ holds $g_\nu \in L_\nu = [o, q_\nu] \subset B_\nu'$.

If $y = e \in G$, then $e_\nu = e \wedge q_\nu \in G_\nu = [o, q_\nu] \subset G$. G_ν is irreducible, i.e. the center Z_ν of G_ν consists of the two elements 0 and q_ν only. It is $\|x\| = \sum_\nu \|x_\nu\|$, $\|y\| = \sup_\nu \|y_\nu\|$ and $\mu(x, y) = \sum_\nu \mu(x_\nu, y_\nu)$. For a denumerable set of base norm spaces B_ν with their duals B_ν' one can construct a space $B = \sum_\nu B$ with dual $B' = \sum_\nu B_\nu'$ by

$$x = (x_1, x_2, \ldots) \in B \text{ if } \sum_\nu \|x_\nu\| < \infty$$

and $$y = (y_1, y_2, \ldots) \in B' \text{ if } \sup_\nu \|y_\nu\| < \infty$$

with $\mu(x, y) = \sum_\nu \mu(x_\nu, y_\nu)$. The norm in B is defined by $\|x\| = \sum_\nu \|x_\nu\|$ and in B' by $\|y\| = \sup_\nu \|y_\nu\|$.

This decomposition of B, B' with respect to the atoms $q_\nu \in Z$ reduces the representation problem to the representation of the dual pairs B_ν, B'_ν.

Therefore we drop the indices ν and consider a pair of dual spaces B, B' with the axioms AV 1.1, AV 1.2s, AVid, AV 3, AV 4 and G irreducible.

Then the following theorems can be proved:

Theorem 6.2 *Each $e \in G$ is the sum of a set of denumerable many orthogonal atoms.*

Theorem 6.3 *If $K_1(e)$ is finite dimensional then e is the sum of finitely many atoms. If in addition AV 4s is valid and if e is the sum of finite many atoms then $K_1(e)$ is finite dimensional and $[0, e]$ (as a subset of G) is an atomic, complete, orthocomplemented and modular lattice.*

With the assumptions AV 1.1, AV 1.2s, AVid, AV 3, AV 4s and G irreducible we have:

Theorem 6.4 *If there are in G at least 4 orthogonal atoms then G is isomorphic to the lattice of the closed subspaces of an Hilbertian space V over a field D.*

Proof: We will show that the presuppositions of Theorem 7.40 in [9] are fulfilled: G is complete; G is atomic; if e is the sum of finitely many atoms then $[0, e]$ is modular (Theorem 6.3); and Theorem 5.36. ■

We do not know whether each lattice G of the closed subspaces of a Hilbertian space can be imbedded in the dual B' of a base norm space B so that (B, B') fulfil all the axioms AV 1.1, AV 1.2s, AVid, AV 3 and AV 4s. This question is of interest because we will see below that under the additional assumption of AV 2f the field D is one of the three fields $\mathbb{R}$, $\mathbb{C}$ or $\mathbb{Q}$.

Now, assuming also axiom AV 2f, we get the following theorems:

Theorem 6.5 *If e is finite (i.e. according to Theorem 6.3: if $K_1(e)$ is finite dimensional or if e is the sum of finite many atoms) then the set $A(e)$ of all atoms $p \leqslant e$ is $\sigma(B', B)$-compact in B'.*

A finite dimensional convex set C is called strictly convex iff the boundary C is equal to $\partial_e C$. If C is strictly convex then C is homeomorphic to a ball.

Theorem 6.6 *If p_1, p_2 are two different atoms of G then $e = p_1 \vee p_2$ is finite and $K_1(e)$ is strictly convex.*

Theorem 6.7 *If e is finite then K_1 is a homeomorphic mapping of $A(e)$ onto $\partial_e K_1(e)$ relative to the $\sigma(B', B)$-topology in B' and the normtopology in B, respectively.*

Theorem 6.8 *For each $e \in G$ $A(e)$ is connected in the $\sigma(B', B)$-topology.*

Theorem 6.9 *If $e \in G$ is finite then $[0, e] \subset G$ is $\sigma(B', B)$-compact.*

Theorem 6.10 *If e is finite then the set $[0, e]_d$ of all elements $e' \in [0, e]$*

with $d(e') = d$ is $\sigma(B', B)$-compact. ($d(e)$ is the number of orthogonal atoms p_ν with $\sum_\nu p_\nu = e$.)

The $\sigma(B', B)$-topology on L can be induced by a norm, which will be denoted $\|\ldots\|_\sigma$. Theorem 5.19 says that the lattice operations are continuous in this norm topology. Using theorems proved in the literature ([9], [10], [11], [12]) from the proceeding theorems results:

Theorem 6.11 *The field D of the Hilbertian space (in Theorem 6.4) is topological and locally compact.*

Using Theorem 6.8 we get

Theorem 6.12 *The field D is one of the three fields* $\mathbb{R}$, $\mathbb{C}$ or $\mathbb{Q}$.

With the help of a theorem proved in [13], the involutory antiautomorphism θ of D turns out to be continuous. Therefore, in the case $D = \mathbb{C}$ we have $\theta\alpha = \bar{\alpha}$ with $\bar{\alpha}$ as the complex conjugate of α.

So the representation theorem is proved:

Theorem 6.13 *If G contains at least 4 orthogonal atoms then G is isomorphic to the lattice of the topologically closed subspaces of a Hilbertspace $\mathcal{H}$ over a field $\mathbb{R}$ or $\mathbb{C}$ or $\mathbb{Q}$.*

This isomorphism can be extended to an isomorphism of B with the space of all self-adjoint operators of the trace class and of B' with the space of all self-adjoint bounded operators where the canonical bilinear form is given by $\mu(x, y) = \mathrm{tr}(x\,y)$. K can be identified with the set of all positive self-adjoint operators w with $\mathrm{tr}(w) = 1$; and L can be identified with all self-adjoint operators g with $0 \leqslant g \leqslant 1$. G is then the set of all projection operators. Commensurability of decision effects turns out to be equivalent with commutativity.

Thanks are due to Professor H. Neumann for critical reading of the manuscript and numerous improvements in the text.

References

[1] G. Ludwig, Grundstrukturen einer physikalischen Theorie (Springer, 1978).
[2] G. Ludwig, The Objective Description of Preparation and Registration as a Basis of Quantum Theory (to appear in Nuovo Cimento).
[3] G. Ludwig, Foundation of Quantum Mechanics, Vol. 1 (Springer Verlag; the translation in English in preparation).
[4] G. Ludwig, A Theoretical Description of Single Microsystems, in: W. C. Price, S. S. Chissick (Ed.), The Uncertainty Principle and Foundations of Quantum Mechanics (John Wiley, 1977) (Correction: APS 4.2 on page 196 has to be dropped.)
[5] G. Ludwig, Einführung in die Theoretische Physik (4 vol.) (Vieweg, 1974–1979).
[6] G. Ludwig, Deutung des Begriffs „Physikalische Theorie" und axiomatische

Grundlegung der Hilbertraumstruktur der Quantenmechanik durch Hauptsätze des Messens, Lecture Notes in Physics **4**, (Springer, 1970).

[7] G. Ludwig, An Axiomatic Basis of Quantum Mechanics (in preparation; Springer Verlag).

[8] G. Ludwig, The measuring process and an axiomatic foundation of quantum mechanics, in: B. D'Espagnat (Ed.), Foundation of Quantum Mechanics (Academic Press, New York, 1971).
G. Ludwig, A physical interpretation of an axiom within an axiomatic approach to quantum mechanics and a new formulation of this axiom as a general covering condition, Notes in Math. Phys. **1** (Marburg, 1971).
G. Ludwig, Meß- und Präparierprozesse, Notes in Math. Phys. (Marburg, 1972).
G. Ludwig, An improved formulation of some theorems and axioms in the axiomatic foundation of the Hilbert space structure of quantum mechancis, Comm. Math. Phys. **26**, 78–86 (1972).
G. Ludwig, Measuring and Preparing Processes, in: A. Hartkämper, H. Neumann (Ed.) Foundations of Quantum Mechanics and Ordered Linear Spaces, Lecture Notes in Physics, **29** (Springer, 1974)
P. Stolz, Attempt of an axiomatic approach of quantum mechanics and more general theories IV, Commun. Math. Phys. **23**, 117–126 (1971).

[9] V. S. Varadarajan, Geometry of Quantum Theory, vol. 1 (v. Norstrand Comp., 1968).

[10] N. Zierler, Axioms for Non-Relativistic Quantum Mechanics, Pac. J. of Math. **11**, 1151–1169 (1961).

[11] M. D. Mac Laren, Atomic Orthocomplemented Lattices, Pac. J. of Math. **14**, 597–612 (1964)

[12] M. D. Mac Laren, Notes on Axioms for Quantum Mechanics, ANL-7065, Argonne Nat. Lab.

[13] N. Zierler, On the Lattice of Closed Subspaces of Hilbert Space, Pac. J. of Math. **19**, 583–586 (1966).

The Concepts of Truth, Possibility and Probability in the Language of Quantum Physics

Peter Mittelstaedt

The language of quantum physics is investigated here with respect to its basic semantic concepts: (1) truth and falsity; (2) necessity and possibility and (3) probability. It is shown in this paper that a universal proof-theoretical semantics, the *extended quantum dialog semantics*, can be constructed, which is equally applicable to the concept of truth, to modalities and to the probability concept. This semantics is found to differ essentially from the traditional semantics of the language of classical physics, i.e. the realistic and metalinguistic semantics of truth (Aristotle, Tarski), the possible worlds semantics of the modalities (Leibniz, Kripke) and the set-theoretical semantics of probability (Kolmogoroff).

Introduction

In recent years the language of quantum physics (Q-language) has been developed in detail by means of the material quantum dialog game [1, 2]. Many formal structures of this Q-language, e.g. the formal object propositional logic [1, 3] (quantum logic), the modal logic of Q-language [4] and the probability calculus for propositions of Q-language [5], – have been shown to differ essentially from the corresponding structures of the language of classical physics (C-language). The discussion of these formal structures can be found in the literature and will hence not be repeated here.

In this paper we are concerned rather with basic semantic concepts of Q-language, the concepts of truth and falsity (§1), the modalities "necessary" and "possible" (§2), and the concept of probability (§3). It will be shown that for the concept of truth a proof-theoretic semantics can be established by means of the material quantum dialog game. This semantics will be called here the *quantum dialog semantics.* In § 2 this dialog-semantics is extended in order to construct a semantics for the modal concepts of necessity and possibility. The modal semantics which is obtained in this way will be called the *extended quantum dialog semantics.* It turns out that this extended quantum dialog semantics can also be applied to the concept of probability in Q-language which leads to a relative-frequency interpretation of quantum-probability.

The universal semantics which is established here for the Q-language should be compared with the semantics usually applied to the concept of truth, the modalities and the probability concept within the framework of the C-language. It turns out that the extended dialog semantics differs essentially from the traditional semantics of truth according to Tarski from the extension of the Tarski semantics to modalities according to Kripke and from the set-theoretical semantics for probability by Kolmogoroff. The reason for these discrepancies lies in the fact that the general concept of reality, which is tacitly presupposed in the C-language-semantics mentioned, is not compatible with the quantum physical reality. The classical reality concept contains some contingent features which justify the traditional semantics, but which are not found to be present in the most general experience of quantum physics. A detailed comparison of the classical and the quantum reality concepts is given in sections 1.3 and 2.3 of this paper.

§ I The concept of truth in Q-language

1) Material propositions

The concept of truth will be considered here within the framework of the propositional object language of quantum physics. In this Q-language we make use of a proof-theoretical concept of truth such that the truth of a given proposition A is established if some convenient proof-procedure $\pi(A)$ leads to a situation $\sigma(A)$, which indicates the success of the performed procedure $\pi(A)$. At this stage of the discussion it is still open whether the procedure $\pi(A)$ is a material measuring process, a purely intellectual argumentation or something in between, which by the community of speakers of the Q-language is accepted as a proof-procedure for the proposition A.

It is important to note that the success of $\pi(A)$ will be considered here as completely sufficient in order to guarantee the truth of A and no other (more realistic) criterion will be taken into account. The reason for this careful restriction will become evident if we compare the present situation with the corresponding problem in the language of classical physics (C-language). C-language is related to classical reality in such a way that it is always possible to refer to this reality as an ultimate criterion for the truth of a given proposition.

We start from a quantum mechanical system S (atom, nucleus, elementary particle), the properties of which can be tested by experiment. Here the physical system S is introduced as an undefined primitive concept, which will not be further explained. In terms of quantum theory a system S is given by some Hilbert space $\mathcal{H}(S)$, which characterizes the dynamical degrees of freedom of S. – In the next step we consider *elementary propositions a, b, c, . . .* which assert that the system S has a certain measurable property. Consequently the proof-procedure $\pi(a)$ for an elementary proposition a is given by a measuring process $M(a) \triangleq \pi(a)$ and the truth of a can be defined as follows:

proposition a is true in the system S	$\curvearrowleft\!\!\curvearrowright$	the measuring process $M(a)$ is successful in S, denoted here by: $S \vdash_M a$.

Here it is required that the success of $M(a)$ is sufficient *and* necessary for the truth of a. If $M(a)$ is not successful, proposition a will be said to be *false*. For elementary propositions we will assume that it can always be decided whether $M(a)$ is successful or not. Hence elementary propositions are *value-definite*, i.e. they are either true or false. The set of elementary propositions will be denoted here by $S_e = \{a, b, c, \ldots\}$.

In quantum theory it is well known that an arbitrary subset $S_e^{(n)} \subseteq S_e$ of n elementary propositions cannot be tested simultaneously on the system S by measurements. After the experimental test of the proposition $a_k^{(n)}$, say, the results of the preceding tests of propositions $a_1^{(n)}, \ldots a_{k-1}^{(n)}$ are generally no longer available. This is only the case if the elements of $S_e^{(n)}$ are pairwise commensurable. In order to incorporate the possibility of pairs of propositions which are not commensurable into the scientific language – without any explicite reference to quantum theory – we introduce the "availability propositions" $k(a, b)$ (commensurability of a and b) and $\bar{k}(a, b)$ (incommensurability) by the following prescription:

proposition $k(a, b)$ is *true* in the system S	$\curvearrowleft\!\!\curvearrowright$	propositions a and b can be tested by measurements on S in an arbitrary sequence without thereby influencing the result of the measurements: $S \vdash k(a, b)$
proposition $\bar{k}(a, b)$ is *true* in the system S	$\curvearrowleft\!\!\curvearrowright$	the result of an experimental test of the one proposition can be changed by an experimental test of the other proposition: $S \vdash \bar{k}(a, b)$

If the conditions on the right hand side are not fulfilled, the propositions $k(a, b)$ and $\bar{k}(a, b)$ respectively will be said to be *false*. It follows from these criteria of truth that $k(a, b)$ is false if and only if $\bar{k}(a, b)$ is true, and that $\bar{k}(a, b)$ is false if and only if $k(a, b)$ is true. However, for availability propositions there is no finite testing-procedure, which decides whether $k(a, b)$, say, is true or false. Hence the criteria of truth for $k(a, b)$ and $\bar{k}(a, b)$ do not imply the value definiteness of the availability propositions.[1] One can, however, by an additional assumption postulate in a consistent manner that for value definite elementary propositions a and b the availability propositions are value definite, as well.[2] Here we will make use of this possibility and postulate the value definiteness of the

[1] According to the truth criteria for $k(a, b)$ and $\bar{k}(a, b)$ the proof procedures for these propositions consist of infinite series of experimental tests. Hence $k(a, b)$ and $\bar{k}(a, b)$ can be expressed by means of quantifiers. It is well known that the value definiteness of elementary propositions is not inherited by the quantifiers. Cf. ref. [6], p. 251.

[2] Cf. ref. [1], p. 123 ff. – The proof that this assumption does not lead to inconsistencies can be found in ref. [7], p. 67.

availability propositions.[3] Elementary propositions and availability propositions are to be proved or disproved by a measurement or at least by a series of measurements, i.e. by material processes. Hence we will refer to these two kinds of propositions as *material propositions.*

2) The material dialog game

Starting from the set $S_e = \{a, b, c, \ldots\}$ of elementary propositions and the availability propositions $k(a, b)$ and $\bar{k}(a, b)$ we further extend our scientific language by incorporating the logical connectives. The logical connectives $a \wedge b$ (a and b), $a \vee b$ (a or b), $a \rightarrow b$ (if a then b) and $\neg a$ (not a) will be defined by means of dialogs. A dialog is a formalized kind of discussion between two participants: the proponent P who asserts a certain proposition and the opponent O who attempts to refute it. The concept of dialog can be defined by general frame rules which read as follows:

F1: At the beginning of the dialog the proponent P asserts the initial argument. In this way the initial position of the dialog is established.

F2: The opponent O has the right to attack the initial argument. The proponent is then committed to defend his argument against this attack. The further dialog consists of a series of arguments which are stated in turn by the two participants P and Q.

F3: With the exception of the initial argument, each argument of a participant either attacks a previous argument of the other participant or defends his own argument against such an attack but is not used simultaneously in both ways.

F4: If one of the participants has no argument to continue, he loses the dialog. In this case the other one wins and the final position of the dialog is established.

In addition to the frame rules, the logical connectives can then be defined by the possibilities of attack and defence in a dialog. In this way one obtains the following attack and defence scheme:

connective		attack	defence
$A \wedge B$	"A and B"	1? 2? $k(A, B)$?	A , B $k(A, B)$!
$A \vee B$	"A or B"	?	A , B $k(A, B)$!
$A \rightarrow B$	"If A then B"	A $k(A, B)$?	B $k(A, B)$!
$\neg A$	"not A"	A	

[3] In ref. [8], a more operational justification for this postulate is given.

Here we have denoted propositions by capital letters A, B, . . . , by $k(A, B)?$ the challenge to prove $k(A, B)$ and by $k(A, B)!$ the successful proof of $k(A, B)$. In contrast to the material propositions which must be proved by material processes, the logical connectives must be proved by dialogs. Hence they will be said to be *dialog definite.*

Using the logical connectives, we can now extend the scientific language – which up to now consists of elementary and availability propositions – by incorporating arbitrary iterations of the logical connectives and availability propositions by the following inductive scheme for the set $\mathcal{S} = \{A, B, \ldots\}$ of propositions:

I) Elementary propositions $A \in \mathcal{S}_e$ are propositions of $\mathcal{S}$

II) If $A \in \mathcal{S}$ then $\neg A \in \mathcal{S}$

III) If $A, B \in \mathcal{S}$ then $A \wedge B, A \vee B, A \rightarrow B$, $k(A, B)$ and $\bar{k}(A, B)$ are propositions of $\mathcal{S}$.

Here we have made use of an obvious extension of the definition of the availability propositions to arbitrary propositions $A, B \in \mathcal{S}$. Instead of a series of material proofs, for the proof of the availability propositions $k(A, B)$ and $\bar{k}(A, B)$ a series of dialogs for propositions A and B must be considered. Since this proof procedure for the availability propositions is not itself a dialog in the strict sense of the frame rules F1–F4, proposition $k(A, B)$ and $\bar{k}(A, B)$ are always considered as material propositions, the proof of which has to pe performed outside of the dialog.

The set $\mathcal{S} = \{A, B, C, \ldots\}$ of propositions which is generated by the inductive scheme mentioned above will be designated here by $\mathcal{S} = \{\mathcal{S}_e; \wedge, \vee, \rightarrow, \neg, k(\ ,\ ), \bar{k}(\ ,\ )\}$. The proof procedure for propositions $A \in \mathcal{S}$ are summarized in the *material dialog game* D_m. The game is called *material*, since the dialogic proof of a proposition A consists apart from dialogic techniques also of the material proofs of elementary and availability propositions which are contained in A.

In order to apply the rules for propositions $A \in \mathcal{S}$ adequately to the material dialog game D_m, one has to procede according to the following *argument rule:*

(A) Every assertion of a partial proposition is the initial argument of a sub-dialog about this partial proposition. If this proposition was used as an attack, the respective obligation to defend will be suspended up to the end of the sub-dialog.

Since the proof of an arbitrary compound proposition $A \in \mathcal{S}$ has to be performed in a D_m-dialog, we define the truth of proposition A as follows:

proposition A is true in the system S	$\curvearrowleft\!\!\curvearrowright$	the D_m-dialog about A in the system S is successful; denoted here by $S \underset{D_m}{\vdash} A$

This explication of the concept of truth for dialog definite propositions $A \in S$ will be called the *quantum dialog semantics.* According to the rules of the material dialog game D_m the truth of a given proposition A depends partly on dialogic strategies and partly on the (non-dialogical) truth of the material propositions contained in A. Consequently the material dialog about a finitely compound proposition $A \in S$ consists of a finite number of dialogic steps, the last one which decides whether P wins the dialog (in which case A is said to be true) or whether P loses (and O wins).the dialog. Hence there is a finite decision procedure for success or loss of a material dialog about a proposition $A \in S$. If P loses the D_m-dialog about A, the proposition A is said to be *false.*

The truth and the falsity of a given proposition $A \in S$ with respect to a system S thus depends on three independent constituents: (1) the properties of the system S, sometimes called its preparation, (2) the truth and falsity of the material propositions contained in A and (3) the strategic possibilities in the dialog. Those propositions $A \in S$ which can be defended successfully in a dialog irrespective of the properties of S and irrespective of the truth of the material propositions contained in A, i.e. by purely formal argumentation, are called *formally true.* The totality of formally true propositions constitute the *formal quantum logic*, which will, however, not be considered here. Here we are concerned rather with the more general concept of the material truth of arbitrary propositions $A \in S$.

3) Dialogic and realistic concepts of truth

According to the quantum dialog semantics the truth of a proposition $A \in S$ relative to the system S depends on the success of a material dialog about A in S. It is not stated here that the success of a D_m-dialog is related to some kind of reality such that the properties of this reality could serve as an alternative criterion of truth. In the spirit of the proof-theoretic semantics used here we have assumed rather that the truth of a proposition $A \in S$ depends exclusely on the success of the corresponding proof-procedure $\pi(A)$ which in the framework of the D_m-semantics consists partly of strategies of argumentation and partly of material measuring processes. In fact, if we consider the physical system S as the reality taken into account, then there is no guarantee that the properties of this system S and the material propositions contained in A fully determine the success of the D_m-dialog about A. The success could depend rather on contingent measuring results and on the strategic possibilities of the participants in the dialog. – There is, however, a special case: the C-language, for which the success of a material dialog is strictly related to the properties of a certain concept of reality. In order to illustrate this possibility and to find out the reason for this particular behaviour, we will briefly recall some important features of C-language.

The construction of C-language can be performed quite analogously to Q-language. Here the physical system S is given by a macroscopic physical system (– in terms of theoretical mechanics by some phase space –) and by elementary propositions $a_1, a_2, \ldots$ which correspond to possible properties of S. The truth (or falsity) of elementary propositions is again defined by a proof (or disproof) due to a measuring process. Elementary propositions will be assumed to be value definite. In contrast to the situation in Q-language there is, however, no restriction for the simultaneous measurement of two propositions $a_i, a_k \in S_e$. Hence the availability propositions $k(a_i, a_k)$ are always true and the propositions $\bar{k}(a_i, a_k)$ are always false. Starting from the set S_e the logical connectives $\wedge, \vee, \rightarrow, \neg$ can then again be defined by the above mentioned possibilities of attack and defence in a dialog - but without the possibilities which correspond to $\bar{k}(a, b)$ and $k(a, b)$. Incorporating the connectives into the language by induction, one arrives at the set $S = (S_e; \wedge, \vee, \rightarrow, \neg)$ of propositions.

If the C-language S is generated by a set $S_e^{(n)} = \{a_1, a_2, \ldots, a_k\}$ of n elementary propositions, one can form $N = 2^n$ propositions $C^{(\nu)} = \alpha_1^{(\nu)} \wedge \alpha_2^{(\nu)} \ldots \alpha_n^{(\nu)}$ (with $\nu = 1, 2, \ldots N$), where $\alpha_k^{(\nu)}$ is either a_k or its counterproposition $\bar{a}_k$ (which is true iff a_k is false). The propositions $C^{(\nu)}$ characterize the possible *contingent situations* of the system S. The $C^{(\nu)}$ are "*perfect*" descriptions of S, since for a given $C^{(\nu)}$ for any proposition $A \in S$ it is completely determined whether A is true or false. Hence, if the preparation of the system is given by a perfect proposition $C^{(\nu)}$, the truth or falsity of any other proposition $A \in S$ follows from the preparation.

In C-language the dialogic concept of truth is equivalent to a realistic concept of truth. In fact, C-language mirrors a situation which is well known from the concept of reality in classical physics (C-reality): A physical system S, which is considered as a sample of C-reality, is characterized by some permanent features and by the totality of contingent properties $C^{(\nu)}$, which can be tested by experimental methods.[4] If the contingent situation $C^{(\nu)}$ of S is given,[5] all other properties of the system are then completely determined.

Hence we find that a proposition $A \in S$ is true in S in the sense of the dialog semantics, i.e. $S \underset{\mathrm{D_m}}{\vdash} A$, if and only if the real situation $\sigma(A)$ which corresponds to the proposition A is the case in the system S. The truth (or falsity) of a proposition $A \in S$ can thus also be defined by the following criterion which refers to the classical reality:

---

4 The concept of reality was first formulated by Kant: "Every thing . . . is likewise subject to the principle of complete determination, according to which if all the possible predicates of things be taken together with their contrary opposites, then one of each pair of contradictory opposites must belong to it." (Critique of pure reason, B599/600).

5 Perfect propositions $C^{(\nu)}$ which characterize the state of a classical system are considered by L. Wittgenstein, ref. [9], No: 4.26 and by R. Carnap, ref. [10]. Carnap calls the propositions $C^{(\nu)}$ "state descriptions".

proposition A is true in the system S $\curvearrowright$ the situation $\sigma(A)$ is the case in system S

This concept of truth, which relates the truth of a given proposition to the occurrence of some real situation, was first conceived by Aristotle.[6] Since the criterion for truth is the corresponding situation in C-reality, the respective semantics will be called *realistic semantics.* For formal languages Tarski[7] gave a reformulation of this criterion which shows that the truth of an object-linguistic proposition can adequately be expressed only by meta-linguistic means. Furthermore, it turns out that this requirement leads to important limitations of the expressibility in a formal language.

Classical reality may be considered as a set of objects (systems) with well defined properties which completely determine all other possible properties of the objects.[8] For this reason C-reality may be used as a model for the concept of truth in C-language. For Q-language, however, we do not know a class of objects with well defined properties, such that the dialogic truth of a proposition A would correspond to the presence of a certain property $\sigma(A)$ of the object system. (Attempts to construct models for Q-language by means of hidden variables, which fulfill the requirements mentioned at least partially, will not be considered here.) Hence in Q-language the material dialog semantics cannot be replaced equivalently by some realistic semantics.

§ II Necessity and possibility in Q-language

1) The meta-language of Q-language

Object propositions of Q-language can be true or false. The statement of truth of a certain proposition $A \in \mathcal{S}$ is, however, not itself an object proposition but a metalinguistic expression. In this section the Q-language will be extended by incorporating the modalities "necessary" and "possible". Similar to the concept of truth, we will consider here also the modalities as meta-linguistic concepts, using again the material quantum dialog game D_m.[9] In order to perform this program in an adequate way, we will first systematically develop the meta-language of Q-language.

6 Aristotle, Metaphysics Γ 1011b.

7 A. Tarski, ref. [11] gives a criterion which can be expressed as follows: "A" is true if and only if A, where A is any proposition and "A" the name of that proposition. Tarski's famous example for this criterion reads "Snow is white" is true if and only if snow is white.

8 This kind of entity can obviously be represented by a "model" in the sense of the mathematical concept of model.

9 Here we use the metalinguistic approach to modalities. It is, however, possible also to incorporate the modalities into the object language by adding two monadic proposition forming operators for "necessary" and "possible" on propositions. Cf. ref. [12].

In the framework of object Q-language, one considers propositions $A \in \mathcal{S}$ with $\mathcal{S} = \{S_e; \wedge, \vee, \rightarrow, \neg; k(\, , \,), \bar{k}(\, , \,)\}$ and physical systems S with properties (preparation) W. If the system S has the preparation W, we write $S = S(W)$. A proposition A is said to be true with respect to a system $S = S(W)$ with preparation W if and only if $S(W) \vdash_{D_m} A$. This statement is not itself a proposition of $\mathcal{S}$, but rather a "meta-proposition". More precisely we will call $S(W) \vdash_{D_m} A$ an "elementary meta-proposition", the proof of which as to be performed by a material dialog within the object dialog game D_m.

In order to define some other kinds of elementary meta-propositions we imagine the total set of equally prepared systems $S = S(W)$ – which of course belong to the same Hilbert space $\mathcal{H}(S)$. If the success of a material dialog does not depend on the individual system, but obviously only on the preparation, we write $W \vdash_{D_m} A$ and use the following terminology:

proposition A is true with respect to W	$\curvearrowleft\!\!\curvearrowright$	$W \vdash_{D_m} A$

Moreover, if a compound proposition A can successfully be defended in a material dialog – irrespective of the preparation of the system – we write $\vdash_{D_m} A$ and define

proposition A is always true	$\curvearrowleft\!\!\curvearrowright$	$\vdash_{D_m} A$

The truth of A, the truth of A with respect to W and the truth of A with respect to $S(W)$ will in general depend on the material propositions contained in A, their truth or falsity and the contingent factual relations among them.

Remark:

There are, however, some compound propositions, the truth of which can be demonstrated in D_m irrespective of the elementary propositions and the availability propositions contained in it. These propositions are said to be "formally true". The totality of formally true propositions constitutes the calculus of "Quantum logic", the Lindenbaum-Tarski algebra of which is given by an orthomodular lattice. – These formal structures will not be considered here.

In order to elaborate, the meta-language of Q-language further, we introduce a few more terms which prove to be very useful. The elementary meta-proposition $W \vdash_{D_m} A$ is proved by means of a D_m-dialog about A, where proposition W is presupposed as hypothesis by the opponent. According to the rules of the material dialog, we thus have:

$$W \underset{D_m}{\vdash} A \leftrightarrow \underset{D_m}{\vdash} W \rightarrow A$$

If a material implication $A \rightarrow B$ is materially true, the relation of "implication" between A and B is said to hold in this case, – which is denoted here by $A \leqslant B$. Hence we have $A \leqslant B \leftrightarrow \underset{D_m}{\vdash} A \rightarrow B$. Furthermore, we can extend Q-language by the null-place-operations $\vee$ and $\wedge$ defined as:

"true position" $\vee$: $W \underset{D_m}{\vdash} \vee$ holds for all W

"false proposition" $\wedge$: there is no W for which $W \underset{D_m}{\vdash} \wedge$ holds .

The set of elementary meta-propositions $S(W) \underset{D_m}{\vdash} A$, $W \underset{D_m}{\vdash} A$ and $\underset{D_m}{\vdash} A$ is denoted here by $\bar{S}_e$. Since elementary meta-propositions $A, B, \ldots$ of this kind are (meta-)statements about the success of certain material dialogs, the commensurability propositions $k(\bar{A}, \bar{B})$ – with $\bar{A}, \bar{B} \in \bar{S}_e$ – are always true, and the corresponding incommensurability propositions $\bar{k}(\bar{A}, \bar{B})$ are always false. This is similar to the situation in object C-language mentioned above. Hence it is not necessary to further discuss availability propositions of elementary meta-propositions.

Starting from the set $\bar{S}_e$ of meta-propositions, we can define meta-connectives $\bar{\wedge}$ (and) $\underline{\vee}$ (or), $\Rightarrow$ (then), $\neg$ (not) by means of meta-dialogs. The meta-dialog-game $\bar{D}$ is governed by the same frame rules F1–F4 mentioned above as the object dialog game – where the term "argument" here means meta-proposition. However, the argument rules for the meta-connectives are much simpler than in the object Q-language, since the possibilities of attack and defence by means of the availability propositions need not be considered. Hence the attack and defence scheme reads:

meta connective		attack	defence
$\bar{A} \bar{\wedge} \bar{B}$	"and"	1? , 2?	$\bar{A}, \bar{B}$
$\bar{A} \underline{\vee} \bar{B}$	"or"	?	$\bar{A}, \bar{B}$
$\bar{A} \Rightarrow \bar{B}$	"then"	$\bar{A}$	$\bar{B}$
$\neg A$	"not"	$\bar{A}$	

Similar to the object language, we can extend here the set $\bar{S}_e$ of elementary meta-propositions by incorporating arbitrary but finite iterations of the connectives. The inductive scheme for the set $\bar{S} = \{\bar{A}, \bar{B}, \ldots\}$ of meta-propositions reads:

I) Elementary metapropositions $\bar{A} \in \bar{S}_e$ are meta-propositions of $\bar{S}$

II) If $\bar{A}, \bar{B} \in \bar{S}$, then $\bar{A} \,\bar{\wedge}\, \bar{B}, \bar{A} \,\underline{\vee}\, \bar{B}, \bar{A} \Rightarrow \bar{B}, \neg \bar{A}$ are meta-propositions of $\bar{S}$.

For the set S we use the obvious notation

$$\bar{S} = \{\bar{S}_e; \bar{\wedge}, \underline{\vee}, \Rightarrow, \neg\}$$

In the framework of the meta-language $\bar{S}$ one can define the truth of a meta-proposition with respect to the meta-dialog game $\bar{D}$, just as in the object language. In general the truth of a given compound meta-proposition $\bar{A}$ will depend on its composition and on the truth (or falsity) of the elementary meta-propositions contained in it. There are, however, meta-propositions $\bar{A}$ which can be defended successfully in a $\bar{D}$-meta-dialog irrespective of the truth or falsity of the elementary meta-propositions of A and which are said to be *formally true*. The formally true meta-propositions constitute the formal meta-logic of quantum logic which turns out to be equivalent to ordinary effective (intuitionistic) logic.[10)]

2) Modalities in Q-language

A quantum physical system S is characterized by some Hilbert space $\mathcal{H}(S)$. A sufficiently large finite or infinite set of systems S^i which belong to a certain Hilbert space $\mathcal{H}_0$ will be denoted by

$$M_{\mathcal{H}_0} := \{S^i : \mathcal{H}(S^i) = \mathcal{H}_0\}$$

If a system S has the preparation W, we write $S = S(W)$ or inversely $W = W(S)$. Using this terminology we can now define the *truth*, the *necessity* and the *possibility* of a proposition A with respect to a system S with preparation W as meta-linguistic terms in the following waa:

proposition A is true in the system $S(W)$: $\quad V_{S(W)}A \curvearrowleft\!\curvearrowright S(W) \underset{D_m}{\vdash} A$

in accordance with the definition mentioned above. A proposition A will be called *necessary with respect to a system* $S(W)$ if it is true in $S(W)$ but irrespective of the individual systems, i.e. for all systems of $M_{\mathcal{H}}$ with preparation W.

---

[10] Cf. ref. [1], p. 140 and ref. [13].

proposition A is necessary in the system $S(W)$:

$$\Delta_{S(W)} A \leftrightarrow W \vdash_{D_m} A$$

$$\leftrightarrow \bar{\bigwedge_i} S^i(W) \vdash_{D_m} A$$

where we have used the meta-all-quantifier $\bar{\bigwedge_i}$, which runs over all elements of $M_{\mathcal{H}}$ with preparation W.

Remark: The meta-proposition $\Delta_{S(W)} A$ is defined here by $W \vdash_{D_m} A$, i.e. by the requirement that the D_m-dialog about A with respect to $S(W)$ is successful irrespective of the individual system S. Hence, if the set $M_{\mathcal{H}}^{(W)}$ of systems $S \in M_{\mathcal{H}}$ with equal preparation W is given operationally as $\{S^i(W)\}_i$ the meta-proposition $W \vdash_{D_m} A$ can be interpreted as $\bar{\bigwedge_i} S^i(W) \vdash_{D_m} A$. This is however, not the definition of $\Delta_{S(W)} A$ but only its ensemble-theoretic interpretation. There are cases such that the truth of $\Delta_{S(W)} A$ can be demonstrated without any recourse to the set $M_{\mathcal{H}}^{(W)}$ of quantum physical systems.

On the basis of these concepts, the possibility of a proposition A with respect to a system $S(W)$ can be defined explicitely:

proposition A is possible with respect to $S(W)$:

$$\nabla_{S(W)} A \leftrightarrow \bar{\neg}\, \Delta_{S(W)} \neg A$$

A is said to be possible (in $S(W)$) if not-A is not necessary (in $S(W)$). According to the material dialog game D_m and the assumptions concerning the value-definiteness of the material object-propositions (cf. § 1.2), for a proposition $A \in S$ and a given system $S(W)$ it can always be decided within a D_m-dialog whether A or $\neg A$ is true. In terms of the meta-language $\bar{S}$ this *dialogical decidability* can be expressed by

$$\bar{\bigwedge_i} \{S^i(W) \vdash_{D_m} A \;\underline{\vee}\; S^i(W) \vdash_{D_m} \neg A\}$$

with $S^i(W) \in M_{\mathcal{H}}^{(W)}$. Using this formula it follows that

$$\bar{\neg}\, \Delta_{S(W)} \neg A \leftrightarrow \bar{\neg}\, \bar{\bigwedge_i} S^i(W) \vdash_{D_m} \neg A \leftrightarrow \underline{\bigvee_i}\, S^i(W) \vdash_{D_m} A$$

where $\underline{\bigvee_i}$ is the meta-existence-quantifier. Here we have made the assumption that for quantifiers the principle of excluded middle holds. For finite sets $M_{\mathcal{H}}$ this is obvious; in the infinite case the value definiteness can be assumed without contradictions.

Consequently the concept of possibility can be expressed by

$$\nabla_{S(W)}A \curvearrowleftright \neg\,\Delta_{S(W)} \neg A \curvearrowleftright \underset{i}{\underline{\vee}}\, S^i(W) \underset{D_m}{\vdash} A$$

Here it should be clearly distinguished that the first term is the *definiendum*, the second one its *definiens* and the third one its *interpretation.*

The definitions of the concepts of truth, necessity and possibility have the immediate consequence that the metapropositions $\Delta_{S(W)}A$, $V_{S(W)}A$ and $\nabla_{S(W)}A$ generally fulfill the relations

$$\Delta_{S(W)}A \curvearrowright V_{S(W)}A \curvearrowright \nabla_{S(W)}A$$

which are well known from the literature.[11)]

Remark: Modalities $\Delta_{S(W)}A$, $\nabla_{S(W)}A$, etc. generally depend on the propositions W and A. There are, however, some compound modalities e.g.

$$\{\Delta_W A \mathbin{\bar{\wedge}} \Delta_W(A \to B)\} \Rightarrow \Delta_W B$$

which are valid irrespective of W, – and irrespective of A and B. (Here we have dropped the letter S since the validity of these modalities does not depend on the individual system). Another example of this kind is $\Delta_W A \Rightarrow \nabla_W A$. These modalities are called formally true modalities. – The formally true modalities are comprehended in the modal-logic of Quantum logic[12)] which, however, will not be considered here. (Cf. ref. [4] and ref. [15]).

In order to give a clearer acccount of the semantics which has been used here for the concepts of truth, necessity and possibility, we use the following more formal characterization of the modalities:

I) Let $M_{\mathcal{H}_0} = \{S^i : \mathcal{H}(S^i) = \mathcal{H}_0\}$ be a set of systems S^i which belong to a Hilbert space $\mathcal{H}_0$.

II) Let $W = W(S)$ be the preparation of a system $S \in M_{\mathcal{H}_0}$ where in general the preparation is a proposition $W \in \mathcal{S}$. Then we define the binary "equivalence relation"

$$R \subseteq M_{\mathcal{H}_0} \times M_{\mathcal{H}_0}, \text{ denoted by } S \simeq S', \text{ with}$$

$$S^i \simeq S^j \curvearrowleftright W(S^i) = W(S^j)$$

---

[11] O. Becker, ref. [14], calls these relations "Modalgefälle".

[12] The modal logic of quantum logic can be written as a calculus M(Q) which consists of three parts: (1) the formal object quantum logic Q, (2) the formal meta logic L(Q) which is equivalent to the ordinary effective (intuitionistic) logic and (3) the modal part of M(Q) which can be expressed by the three laws

M0(1)	$\Delta \wedge \subseteq \bar{\wedge}$
M0(2)	$\Delta(A \to B) \subseteq \Delta A \Rightarrow \Delta B$
M0(3)	$\Delta A \mathbin{\bar{\wedge}} \Delta B \subseteq \Delta(A \wedge B)$

where the sign $\subseteq$ means the meta-implication.

(Here the sign "=" means the equivalence between propositions. In terms of quantum mechanics this corresponds to the equivalence-relation between subspaces of Hilbert space.)

The interpretation of the modalities mentioned above can now be reformulated in the following way: Let $A \in \mathcal{S}$ and $S \in M_{\mathcal{H}}$; then the concepts of truth, necessity and possibility can be given the following meaning

true: $\mathrm{V}_s A \leftrightarrow S \vdash_{\mathrm{D_m}} A$

necessary: $\Delta_s A \leftrightarrow \bigwedge_i \{S^i \simeq S : S^i \vdash_{\mathrm{D_m}} A\}$

possible: $\nabla_s A \leftrightarrow \bigvee_i \{S^i \simeq S : S^i \vdash_{\mathrm{D_m}} A\}$

The concepts of truth,. necessity and possibility are thus related to a system $S \in M_{\mathcal{H}}$ and a proposition $A \in \mathcal{S}$. However, whereas "true" can be interpreted by a single dialog on an individual system (quantum dialog semantics), "necessary" and "possible" are interpreted with respect to a set $M_{\mathcal{H}}$ of systems and a series of dialogs performed on these systems. This explication will be called "*extended quantum dialog semantics*". The interpretation of the modalities in this way is in the spirit of Leibniz and Kripke [16]. Leibniz explains the term "A is necessary" by "A is true in every possible world", whereas Kripke introduces a binary (accessibility)-relation between possible worlds, and explains "A is necessary" by "A is true in every *accessible* world". – In the present approach the 'possible worlds' are given the set $M_{\mathcal{H}}$ of systems and the accessibility-relation by the equivalence $S \simeq S'$. Hence "A is necessary in S" means "A is true in all $\tilde{S} \in M_{\mathcal{H}}$ with the same preparation as S" and "A is possible in S" means that there is at least one equally prepared system $\tilde{S}$ such that A is true in $\tilde{S}$.

Remark: There is, however, a decisive difference between the semantics of Leibniz and Kirpke which is conceived for C-language and the extended quantum dialog semantics. In C-language "possible" means "true" and "real" at least in one accessible, possible world. According to the extended quantum dialog semantics "possible" means that at least one $\mathrm{D_m}$-dialog performed on the accessible, possible systems (worlds) is successful. If the preparation W and the proposition A are incommensurable, this does not mean that a system $S = S(W)$ possesses the property A, – moreover we know that none of the systems $S^i(W)$ has the property A in an objective sense. (A is not objectifiable in $S(W)$).

Remark:
The set $M_{\mathcal{H}}^{(W)}$ of systems with equivalent preparation W was denoted here by $\{S^i(W)\}_i$. However it is not clear what the index $i \in I$ means in this case, since for maximal propositions W (i.e. 1-dimensional subspaces) the system is completely described and no other number i can be attached to the sys-

tem. In fact quantum systems in the same "state" cannot be distinguished and hence obey Bose or Fermi-statistics. On the other hand, the systems $S(W)$ are produced operationally (e.g. by an accelerator) in time and thus the production time t_i might be used as index-property. (An analogous argument can be put forward for the production place x_i). Obviously this argumentation presupposes space-time translational invariance for a system $S(W)$, or more generally any kind of superselection rule which allows the systems to be individualized. This means that the existence of external symmetries belongs to the necessary preconditions of the extended dialog semantics which is applied here to the modalities.

3) The indispensability of the modalities in Q*-language*

In the last subsection we introduced the modalities as an extension of Q-language and its meta-language. The same method can obviously be applied to C-language [17]. There is, however, a decisive difference between modalities in C-language and in Q-language. Under the premises which were formulated for the C-language in § 1.3, the modalities "necessary" and "possible" can always be completely eliminated, whereas in Q-language the modalities turn out to be indispensible constituents of the scientific language.

The C-language which is considered here is generated by the set $S_e^{(n)} = \{a_1, \ldots a_n\}$ of n elementary, value-definite propositions $a_1, a_2 \ldots$, which correspond to possible properties of a system S. The logical connectives $\wedge, \vee, \rightarrow, \neg$ can then be introduced by means of dialogs, where availability propositions $k(\ ,\ )$ and $\bar{k}(\ ,\ )$ need not be taken into account. If one defines the "implication" $\leqslant$ and the proposition V (truth) and $\mathsf{\Lambda}$ (falsity) analogously as in Q-language, one arrives at the set $S = \{S_e; \wedge, \vee, \rightarrow, \neg, \leqslant, \mathsf{V}, \mathsf{\Lambda}\}$ of propositions. For a proposition $A \in S$, the material truth is defined by $\underset{D_m}{\vdash} A$ and the formal truth by the requirement that the truth of A does not depend on the truth-values of the elementary propositions contained in A. The totality of formally true propositions constitute the calculus of formal C-logic, the Lindenbaum-Tarski-Algebra of which turns out to be a free Boolean lattice $L_B(S_e)$ generated by S_e.

Some important features of C-language will be reported briefly here without proofs. On the basis of $S_e^{(n)}$ one can form $2^n = N$ contingent situations $C^{(\nu)} = \alpha_1^{(\nu)} \wedge \alpha_2^{(\nu)} \ldots \wedge \alpha_n^{(\nu)}$ $(\nu = 1, \ldots N)$ with $\alpha_i^{(\nu)} \in \{a_i, \neg a_i\}$ and $C^{(\nu)} \neq \mathsf{\Lambda}$. The propositions $C^{(\nu)}$ are *perfect* in the sense that for any $A \in S$ a contingent situation $C^{(\nu)}$ implies either A or $\neg A$, or more formally $C^{(\nu)} \leqslant A \veebar C^{(\nu)} \leqslant \neg A$. Moreover the $C^{(\nu)}$ are *maximal* propositions, i.e. $\mathsf{\Lambda} \leqslant X \leqslant C^{(\nu)}$ implies $X = \mathsf{\Lambda} \veebar X = C^{(\nu)}$. In terms of lattice theory this means that the $C^{(\nu)}$ are *atoms* in the lattice $L_B(S_e)$. Furthermore the lattice $L_B(S_e)$ is *atomic*, i.e. for any $A \in L_B(S_e)$ there exists an atom α with $\alpha \leqslant A$, and the atoms are just given by the elements $C^{(\nu)}$.

It is a consequence of these properties that for any proposition W with $W \leqslant A$ there exists a *maximal* proposition W_0 with $W_0 \leqslant W$ such that $W_0 \leqslant A \veebar W_0 \leqslant \neg A$. If the modalities $\Delta_{S(W)}A$ and $\nabla_{S(W)}A$ are replaced by $W \leqslant A$ and $\neg(W \leqslant \neg A)$, respectively, we find that for maximal W_0 it follows

$$\nabla_{S(W_0)}A \leftrightarrow \neg(W_0 \leqslant \neg A) \leftrightarrow W_0 \leqslant A \leftrightarrow \Delta_{S(W_0)}A$$

i.e. the modalities "necessary" and "possible" coincide. Furthermore, for any arbitrary preparation W there exists a "perfect" preparation W_0 such that $W_0 \leqslant W$ and the modalities Δ and ∇ become equivalent with respect to W_0. Hence the modality of possibility is meaningful only as long as the knowledge W about the system S is not perfect. It can, however, always be made perfect by adding new information. Consequently the modality of "possibility" is either meaningless or it expresses an always avoidable subjective ignorance and is thus avoidable as well.

In Q-language, the situation is quite different. If one starts also here with a set $S_e^{(n)} = \{a_1, a_2, \ldots a_n\}$ of n elementary propositions, the language generated by $S_e^{(n)}$ is given by $S = \{S_e; \wedge, \vee, \rightarrow, \neg, k(\ ,\ ), \bar{k}(\ ,\ ), \leqslant, \bigwedge, \bigvee\}$ The formally true propositions are comprehended in the calculus Q of quantum logic, the Lindenbaum-Tarski algebra of which is given by an orthomodular lattice – which in addition will be assumed to satisfy atomicity and the covering law.[13]

Under these assumptions there are atoms (maximal elements) W_0 which have, however, no obvious relations to the elementary propositions and which are not perfect in the sense mentioned above. Consequently even for maximal W_0 the modalities $\Delta_{W_0}A$ and $\nabla_{W_0}A$ will not coincide for arbitary $A \in S$. In contrast to C-language the following weak theorems hold in Q-language:[14]

1) If W_0 is maximal and if $A \in S$ is commensurable with W_0 (denoted here by $W_0 \sim A$), then it holds: $\Delta_{W_0}A \veebar \Delta_{W_0} \neg A$.

2) There are no *perfect* propositions W_0 in S.

Hence in Q-language there is no simple method for eliminating the modalities $\Delta_W A$ and $\nabla_W A$. The recourse to perfect propositions – always possible in C-language – is restricted here to maximal propositions $W_0 \leqslant W$ and such propositions A for which $W_0 \sim A$ holds. Consequently in Q-language we have to distinguish four different situations for the modality of possibility:

---

[13] Cf. ref. [1], p. 16.

[14] For the proofs, cf. ref. [4].

- $\nabla_W A$
  - $W \sim A$
    - $W = W_0 \curvearrowright \nabla_{W_0} A \equiv \Delta_{W_0} A$ meaningless (1)
    - $W \neq W_0 \curvearrowright \nabla_W A$ contains the ignorance case (2)
  - $\neg (W \sim A)$
    - $W \neq W_0 \curvearrowright \nabla_W A$ meaningfull (3)
    - $W = W_0 \curvearrowright \nabla_{W_0} A \not\equiv \Delta_{W_0} A$; A objectively undecided (4)

In Q-language, even for maximal propositions W_0, it happens that $\nabla_{W_0} A$ is different from $\Delta_{W_0} A$, namely if $\neg (W_0 \sim A)$ (case 4). The reason is that – in contrast to C-language – maximal propositions W_0 are not perfect, except with respect to those A commensurable to W_0. Hence if W_0 is maximal and $\neg (W_0 \sim A)$ (case 4) we have $\neg (\Delta_{W_0} A \mathrel{\underline{\vee}} \Delta_{W_0} \neg A)$ and thus $\nabla_{W_0} A \mathrel{\bar{\wedge}} \nabla_{W_0} \neg A$, which means that for a system $S(W_0)$ with preparation W_0 it is *objectively undecided* whether A or $\neg A$ is true. Since W_0 cannot be made perfect by incorporating new information into Q-language, "possibility" is not related to an avoidable subjective ignorance. Therefore the modalities "necessary" and "possible" are indispensable elements of Q-language.

§ III Probability in Q-language

1) Validity functions for modalities

The explanations for the concepts of truth, necessity and possibility which were given in the preceding section, can conveniently be expressed by means of validity functions ν_α. A validity function ν_α, with $\alpha \in \{\mathrm{V}, \Delta, \nabla\}$ is a mapping

$$\nu_\alpha : S \times M_{\mathcal{H}} \rightarrow \{1, 0\}$$

where 1 stands for "true" and 0 for "false" [15] According to the explanations given above, the definitions of the functions $\nu_\alpha(A, S)$ read:

I) Truth

$$\nu_{\mathrm{V}}(A, S) = \begin{cases} 1 & \curvearrowright \quad S \underset{\mathrm{D_m}}{\vdash} A \\ 0 & \curvearrowright \quad \text{otherwise} \end{cases}$$

[15] For this reason some authors use T and F instead of 1 and 0, respectively. Cf. ref. [18], p. 31.

II) Necessity

$$\nu_{\Delta}(A,S) = \begin{cases} 1 & \curvearrowleft\!\!\curvearrowright \quad \bigwedge_i \{S^i \simeq S : S^i \underset{\mathrm{D_m}}{\vdash} A\} \\ 0 & \curvearrowleft\!\!\curvearrowright \quad \text{otherwise} \end{cases}$$

III) Possibility

$$\nu_{\nabla}(A,S) = \begin{cases} 1 & \curvearrowleft\!\!\curvearrowright \quad \bigvee_i \{S^i \simeq S : S^i \underset{\mathrm{D_m}}{\vdash} A\} \\ 0 & \curvearrowleft\!\!\curvearrowright \quad \text{otherwise} \end{cases}$$

Since these functions map all pairs $(A, S) \in \mathcal{S} \times M_{\mathcal{H}}$ into $\{0, 1\}$ the validity functions ν_α are *bivalent valuations* of the language $\mathcal{S}$.

For a material proposition $A \in \mathcal{S}$, the dialogic proof of A reduces to the material proof $S \vdash A$ of A in S. In this way, the valuation $\nu_{\vee}(A, S)$ is defined for all material propositions $A \in \mathcal{S}$. For compound propositions one has to recall the dialogic definitions of the connectives $\wedge, \vee, \rightarrow, \neg$. Furthermore one has to add postulates for $\nu_{\vee}$ which take account of the formal commensurabilities. In this way the validity function $\nu_{\vee}(A, S)$ can be established for all elements of $\mathcal{S}$, which corresponds to a reformulation of the calculus Q of full quantum logic in terms of validity functions. The validity functions $\nu_{\Delta}(A, S)$ and $\nu_{\nabla}(A, S)$ are determined by the definitions (II) and (III) respectively. On the other hand one can deduce from these definitions some relations for the functions $\nu_{\Delta}(A, S)$ and $\nu_{\nabla}(A, S)$ which allow for eliminating the definitions and can thus be used as axioms for ν_{Δ} and ν_{∇}. In this way one finds:

(Δ1) $\nu_{\Delta}(\Lambda, S) = 0$

(Δ2) $\nu_{\Delta}(A, S) = 1$ and $\nu_{\Delta}(B, S) = 1$ implies $\nu_{\Delta}(A \wedge B, S) = 1$

(Δ3) $\nu_{\Delta}(A \rightarrow B, S) = 1$ implies: If $\nu_{\Delta}(A, S) = 1$ then $\nu_{\Delta}(B, S) = 1$

These laws may be considered as a reformulation of the axioms M0(1), M0(2) and M0(3) of the modal logic of quantum logic mentioned above (Note 12 and ref. [15]) in terms of validity functions.

According to the definitions of the functions $\nu_{\Delta}(A, S)$ and $\nu_{\nabla}(A, S)$ for the necessity and possibility respectively, the validity of these modalities does not depend on the individual system S, but rather on the preparation $S = S(W)$ of this system. Hence it will be often more convenient to write

$$\nu_{\nabla}(A, S(W)), \qquad \nu_{\Delta}(A, S(W))$$

in order to express the dependence of the validity functions on the preparation W. We will come back to this more adequate notation in the following subsections.

2) Probability-functions

Probability will be interpreted here as a quantification of the possibility concept. Hence we will introduce the probability in a similar way as was done for the modalities by means of validity functions. Starting from the language $\mathcal{S}$, the set $\mathcal{M}_{\mathcal{H}} = \{S^i\}_i$ of systems S^i with Hilbert-space $\mathcal{H}$ and subsets $\mathcal{M}_{\mathcal{H}}^{(W)} \subseteq \mathcal{M}_{\mathcal{H}}$ of systems with equal preparation W such that $\mathcal{M}_{\mathcal{H}} = \{\mathcal{M}_{\mathcal{H}}^{(W)}\}_W$ holds, a *"probability function" p* is defined as a mapping

$$p : \mathcal{S} \times \{\mathcal{M}_{\mathcal{H}}^{(W)}\}_W \to [0, 1]$$

Let $\mathcal{M}_{\mathcal{H}}^{(W)} = \{S^i(W)\}_i$ be a set of systems with preparation W. (We will assume here that the index i has an operational meaning due to some superselection properties). In order to investigate the truth of a proposition $A \in \mathcal{S}$ with respect to a given system $S(W)$ we have to consider the dialog-position $_{\langle W \rangle}| A$ within the material dialog game $\mathrm{D_m}$. The arguments of O which are available at the beginning of the A-dialog are summarized in the proposition W (the preparation). If the position $_{\langle W \rangle}| A$ in the dialog about the individual system S is a position of success, the material dialog about A can successfully be performed in $S(W)$, i.e. $S(W) \underset{\mathrm{D_m}}{\vdash} A$. As already mentioned above, the material truth of A with respect to $S(W)$ will generally depend on the preparation W and on the individual system $S = S(W)$.

If material dialogs about A are performed on n systems $S^i(W) \in \mathcal{M}_{\mathcal{H}}^{(W)}$ $(i = 1, 2, \ldots n)$, there will be, say, a number $\sigma^{(n)} = \sigma^{(n)}(_{\langle W \rangle}|A)$ of successful dialogs. Hence the relative frequency $f^{(n)}(_{\langle W \rangle}|A)$ of successful dialogs is given by

$$f^{(n)}(_{\langle W \rangle}|A) = \frac{\sigma^{(n)}(_{\langle W \rangle}|A)}{n}$$

On the basis of this relative frequency concept we interpret the probability function $p = p_{\langle W \rangle}(A)$ as the limit of the relative frequency

$$p_{\langle W \rangle}(A) = \lim_{n \to \infty} f^{(n)}(_{\langle W \rangle}|A)$$

Since the relative frequency $f^{(n)}$ depends on $A \in \mathcal{S}$ and on the preparation W the probability function p depends as well on A and W but not on the individual system $S(W)$. In accordance with this fact we have denoted here the probability function as $p_{\langle W \rangle}(A)$ expressing the dependence on A and W.

In order to emphasize the analogy to the modalities and the extended dialog semantics, one can, however, express the probability function also as a function $p_S(A)$ which depends on A and on the individual system S. According to the relative-frequency interpretation the probability $p_S(A)$ for A with respect to the system S is then interpreted as the limit of relative frequency of successful dialogs about A at system S' with $S' \simeq S$, i.e.

$$p_s(A) = \lim_{n \to \infty} \frac{\nu \{S^i \cong S : S^i \underset{D_m}{\vdash} A\}_i}{n}$$

where $\nu \{\alpha^i\}_i$ denotes the number of elements of the set $\{\alpha^i\}_i$, and $n = \nu \{S^i \simeq S\}_i$. The fact that the probability of A with respect to a system S depends actually only on the preparation $W = W(S)$ can then be expressed by the relation

$$p_S(A) = p_{\langle W(S)\rangle}(A) = p_{\langle W\rangle}(A)$$

In this way we have established an interpretation of the probability concept which is based on the same *extended quantum dialog semantics*, which has already been used for the interpretation of the modalities. In fact the probability function $p_S(A)$ is related here to the set $M_{\mathcal{H}}$ of "possible worlds" as a basic ensemble in connection with the "accessibility relation" which selects the elements of $M_{\mathcal{H}}$ with the same preparation as S. Hence the probability $p_S(A)$ can be interpreted as: "the limit of the relative frequency of successful A-dialogs on possible systems (worlds) which are equally prepared as S (accessible for S)".

In a manner similar to the preceding section (§2.3) one can argue that in contrast to C-language in Q-language the probability concept is an indispensable element of the scientific language. In fact in C-language the probability function $p_{\langle W\rangle}(A)$ differs from the values 1 (true) and 0 (false) only because of the subjective ignorance which is contained in the imperfect knowledge W about the system $S(W)$. This knowledge can, however, always pe perfected by incorporating new information about S. For a perfect proposition W_0 with $W_0 \leqslant W$ the probability $p_{\langle W_0\rangle}(A)$ will be either 1 or 0, i.e. the proposition A is either true or false. Hence the probability concept can always be eliminated in C-language. In Q-language the situation is quite different. There are no perfect propositions and for maximal propositions W_0 (which correspond to one dimensional state vectors) the probability function $p_{\langle W_0\rangle}(A)$ can, in general, take any value $0 \leqslant p_{\langle W_0\rangle}(A) \leqslant 1$. In Q-language the probability concept expresses the "objective undecidedness" of A with respect to W_0 and cannot be eliminated by recourse to some perfect knowledge.

3) The probability calculus

The numerical value of a probability function $p_{\langle W\rangle}(A)$ generally depends on the compound propositions W and A. The calculation of this value can be performed by means of a *probability calculus* $\mathcal{P}$ which represents a twofold recoursive scheme with respect to the propositions W and A. Furthermore this calculus allows for the derivation of relations between two or more probability functions, e.g. $p_{\langle W \rangle}(A)$ and $p_{\langle W' \rangle}(A \wedge B)$ which hold irrespective of the preparations $W, W', \ldots$ and of the propositions $A, B, \ldots$ and which will be called *formally true probability* relations. An example of this kind is the inequality $p_{\langle W \rangle}(A \wedge B) \leqslant p_{\langle W \rangle}(A)$. According to the extended dialog semantics and the relative frequency interpretation, the probability-calculus can be established in the following way:

α) Elementary propositions and availability propositions.
The proof procedure $\pi(A)$ for a material proposition A is given by a measuring process or by a series of measuring process. If the proof procedure for A with respect to the system $S(W)$ is successful, we write $S(W) \vdash A$. Let n be the number of proof attempts at systems $S^i(W) \in \mathcal{M}_{\mathcal{H}}^{(W)}$ The number of successful proofs will then be denoted by $\sigma^{(n)}(W \vdash A)$ and the relative frequency $f^{(n)}$ of successful proofs by $f^{(n)}(W \vdash A)$. A proposition A will be called *probability definite*, if for all preparations W there exists a uniquely defined limit of the relative frequency $f^{(n)}(W \vdash A)$. The quantum probability calculus which is established here is based on the *hypothesis* that material propositions are probability definite. It is obvious that a proof of this hypothesis cannot be given within the framework of the present approach.[16]

β) Compound propositions.
In order to determine probabilities for dialog definite compound propositions $A \in \mathcal{S}$, we use the extended dialog semantics and recall the dialogic definitions of the logical connectives. In this way one obtains for instance the reduction formula for the conjunction

$$p_{\langle W \rangle}(A \wedge B) = p_{\langle W \rangle}(k(A, B))\, p_{\langle W, k(A,B) \rangle}(A)\, p_{\langle W, k(A,B), A \rangle}(B)$$

where $\langle W, k(A, B) \rangle$ denotes a preparation which is obtained by a successful proof of $k(A, B)$ on a system S with preparation W. Furthermore we mention the rules

$$\mathsf{V} \leqslant W \to A \quad \curvearrowleft\!\!\curvearrowright p_{\langle W \rangle}(A) = 1$$
$$\mathsf{V} \leqslant W \to \neg A \curvearrowleft\!\!\curvearrowright p_W(A) = 0$$

and the relation

[16] There are, however, strong indications that the hypothesis is correct. In the framework of quantum mechanics, a quantum mechanical "law of large numbers", which justifies the hypothesis mentioned can be shown to be valid. This result may obviously be considered as a proof of the hypothesis only with respect to the self-consistency of language and theory under consideration. (Cf. ref. [19] and [20]).

$$p_{\langle W \rangle}(\neg A) = 1 - p_W(A)$$

which follows from the definition of the negation.

The probability calculus $\mathcal{P}$ which can be established in this way is a twofold recursive scheme for the calculation of probability functions $p_{\langle W \rangle}(A)$ of compound propositions W and A. Since material propositions are probability definite, it is a consequence of this probability calculus that all finitely compound propositions $A \in \mathcal{S}$ are probability-definite as well. The probability calculus $\mathcal{P}$ can be found in the literature [5] and will hence not be presented here.

Starting from the calculus $\mathcal{P}$ the following relations between probability, functions with equal preparation can easily be shown to be valid for arbitrary finitely compound preparations W

p1) $0 \leqslant p_{\langle W \rangle}(A) \leqslant 1$

p2) $p_{\langle W \rangle}(\Lambda) = 0, \quad p_{\langle W \rangle}(\mathrm{V}) = 1$

p3) If A_i is a sequence of propositions $A_i \in \mathcal{S}$ which satisfy $A_i \leqslant \neg A_j (i \neq j)$ then

$$p_{\langle W \rangle}(A_1 \vee A_2 \vee \ldots) = \sum_i p_{\langle W \rangle}(A_i)$$

p4) $p_{\langle W \rangle}(A) = 1$ and $p_{\langle W \rangle}(B) = 1$ implies $p_{\langle W \rangle}(A \wedge B) = 1$

The laws p1)–p4) are well known as formal axioms of abstract Q-probability, in which $A, B, \ldots$ are elements of an orthomodular lattice L_q [2], 22]. The axioms of Q-probability are slightly different from the Kolmogoroff axioms which are related to the elements of a Boolean lattice L_B instead of L_q.

4) Probability and possibility

According to the relative frequency interpretation of probability, the function $p_S(A)$ may be considered as a refinement of the validity function $\nu_\nabla(A, S)$ of the possibility. In fact, the validity functions $\nu_\nabla(A, S)$ and $\nu_\Delta(A, S)$ for the modalities "possible" and "necessary" can be expressed by the possibility function:

$$\nu_\nabla(A, S) = \begin{cases} 1 & \text{iff } p_S(A) \neq 0 \\ 0 & \text{iff } p_S(A) = 0 \end{cases}$$

$$\nu_\Delta(A, S) = \begin{cases} 1 & \text{iff } p_S(A) = 1 \\ 0 & \text{iff } p_S(A) \neq 1 \end{cases}$$

Hence starting from the axioms (p1)–(p4) of the probability function

$p_S(A)$, it is to be expected that the axioms ($\Delta 1$) – ($\Delta 3$) of the validity function $\nu_\Delta(A, S)$ and the relations for $\nu_\nabla(A, S)$ mentioned above can be derived. This is actually the case. The proofs are as follows:

($\Delta 1$): From (p2) we get $p_s(\Lambda) \neq 1$ and thus $\nu_\Delta(\Lambda, S) = 0$.

($\Delta 2$): $\nu_\Delta(A, S) = 1$ and $\nu_\Delta(B, S) = 1$ imply $p_s(A) = p_s(B) = 1$. From (p4) it then follows $p_s(A \wedge B) = 1$ and hence $\nu_\Delta(A \wedge B, S) = 1$.

($\Delta 3$): lemma α : From (p3) we get $p_s(A) + p_s(\neg A) = p_s(A \vee \neg A) = 1$. lemma β: Using (p3) $A \leqslant C$ implies $p_s(A) + p_s(\neg C) = p(A \vee \neg C)$ and due to lemma (α) and (p1) it follows $p_s(A) \leqslant p_s(C)$. Proof of ($\Delta 3$): $\nu_\Delta(A \to B, S) = 1$ and $\nu_\Delta(A, S) = 1$ implies $p_s(A \to B) = p_s(A) = 1$ and hence using (p4) we get $p_s(A \wedge (A \to B)) = 1$. Applying lemma ($\beta$) and $A \wedge (A \to B) \leqslant B$ it follows $p_s(B) = 1$ and hence $\nu_\Delta(B, S) = 1$.

Hence we find that the validity functions $\nu_\Delta(A, S)$ and $\nu_\nabla(A, S)$ of the modalities are intimately related to the probability functions. The basic laws ($\Delta 1$), ($\Delta 2$), ($\Delta 3$) for the validity functions of the modalities follow from the axioms (p1)–(p4) of Q-probability, and the validity functions of the modalities express the gross-structure of the respective probability function. For modalities as well as for probabilities the underlying semantics is the same extended quantum dialog semantics. The basic set of "possible words" is given in both cases by the ensemble of systems with the Hilbert space $\mathcal{H}$ and the "accessibility relation" R is given by the equivalence relation between equally prepared systems.

Remark: It should be emphasized that according to the extended quantum dialog semantics there is again a decisive difference between the interpretation of the probability in C-language and in Q-language – similar as in the case of the modalities. In C-language "relative frequency" $f_W^{(n)}(A)$ means the ratio of the number of systems (with preparation W) which *posses* the property A to the number n of all systems with preparation W. In Q-language $f_W^{(n)}(A)$ means the ratio of the number $\sigma^{(n)}$ of successful A-dialogs at systems $S^i(W)$ to the number n of all performed dialogs. If the preparation W and the proposition A are incommensurable we know that there is no system $S^i(W)$ which *possesses* the property A in an objective sense. For every individual system it is *objectively undecided* whether A is true with respect to W or not.

References

[1] P. Mittelstaedt, Quantum Logic, D. Reidel Publishing Co., Dordrecht, Holland, 1978.

[2] E. W. Stachow, Zur Begründung der Quantenlogik durch die argumentativen Vorbedingungen einer Wissenschaftssprache, in: P. Mittelstaedt, J. Pfarr (eds.), Grundlagen der Quantentheorie, Bibliographisches Institut, Mannheim 1979.

[3] E. W. Stachow, Completeness of Quantum Logic, Journal of Philosophical Logic 5 (1976) 237–280.

[4] P. Mittelstaedt, The Modal Logic of Quantum Logic, Journal of Philosophical Logic 8 (1979) 479–504.

[5] E. W. Stachow, An Operational Approach to Quantum Probability, in: C. Hooker (ed.), Physical Theory as Logico-Operational Structure, p. 285–321, D. Reidel Publishing Co., Dordrecht, Holland, 1978.

[6] K. Lorenz, Die dialogische Rechtfertigung der effektiven Logik, in: F. Kambartel und J. Mittelstraß (eds.), Zum normativen Fundament der Wissenschaft, Athenäum, Frankfurt (1973).

[7] P. Lorenzen, Formal Logic, D. Reidel Publishing Co., Dordrecht, Holland (1965).

[8] P. Mittelstaedt and E. W. Stachow, The Principle of Excluded Middle in Quantum Logic, Journal of Philosophical Logic 7 (1978) 181–208.

[9] L. Wittgenstein, Tracatus logico-philosophicus, London 1922; Basil Blackwell 1959.

[10] R. Carnap, Meaning and Necessity, Chicago (1947).

[11] A. Tarski, "The Concept of Truth in Formalized Languages" in his Logic, Semantics, Metamathematics, J. H. Woodger (trans.) Oxford, Clarendon (1956).

[12] G. E. Hughes and M. J. Cresswell, An Introduction to Modal Logic, London (1968).

[13] P. Mittelstaedt, The Meta-Logic of Quantum Logic, Proceedings of the PSA 1978, P. D. Asquith and I. Hacking (eds.), Vol. 1, p. 249, Philosophy of Science Association, East Lansing, Mich. (1978).

[14] O. Becker, Untersuchungen über den Modalkalkül, Meisenheim, Glan (1952).

[15] F. J. Burghardt, Modale Quantenmetalogik mit Dialogischer Begründung, Dissertation Universität Köln (1979).

[16] S. A. Kripke, A Completeness Theorem in Modal Logic, Journal of Symbolic Logic **24** (1959), 1–14.

[17] P. Lorenzen und O. Schwemmer, Konstruktive Logik, Ethik und Wissenschaftstheorie, Bibliographisches Institut, Mannheim (1977).

[18] Bas C. van Fraassen, Formal Semantics and Logic, The Macmillan Company, New York (1971).

[19] E. W. Stachow, Operational Quantum Probabilities, Proceedings of the International Congress on Logic, Methodology and Philosophy of Science, Hannover (1979).

[20] W. Ochs, The Strong Law of Large Numbers in Quantum Probability Theory, Journal of Philosophical Logic **6** (1977), 473–481.

[21] P. Mittelstaedt, On the Applicability of the Probability Concept of Quantum Theory, in: Harper and Hooker (eds.), Foundations of Probability Theory, Statistical Inference and Statistical Theories of Sicence, Vol. III, 155–167, D. Reidel Publishing Co., Dordrecht, Holland (1976).

[22] J. M. Jauch, The Quantum Probability Calculus, in: J. Leite Lopes and M. Paty (eds.), Quantum Mechanics, a Half Century Later, D. Reidel Publ. Co., Dordrecht 1977, p. 39–62.

The Propositional Language of Quantum Physics

Ernst-Walter Stachow

1. Introduction

The purpose of this contribution is to consider the "quantum logical" approach to quantum mechanics more rigorously from a logican's point of view. Whereas the *mathematical* logic of "quantum logical" systems (the *representation* of logical systems by mathematical structures) has been developed extensively, it seems to us that some questions of *philosophical* logic (concerning the *justification* of "quantum logical" systems) have not yet been investigated in a satisfactory way. Usually the philosophical study of logic starts with an analysis of language. In some cases it is possible to formalize a language in such a way that the formalization comprises a *syntax* (e.g. the set of *sentences*) and a *formal semantics* (e.g. the *interpretations* of the set of sentences). Once such a *formal language* is established one may proceed to *logical systems* which are "appraised" by the language in the sense that they are *complete* and *sound* with respect to the language. In this contribution we restrict our analysis to the *propositional object-language* of quantum physics. Our point of departure is a particular semantics of that language, known as *dialogic semantics*, which was developed in [1]. Within the framework of the dialogic approach, a language is regarded as a formalized manner of argumentation with respect to *propositions.* Thus the structural possibilities of argumentation establish the semantics of the language.

There is another point of view, which is usually found in the standard literature of propositional logic, namely that a language is used for the description of the possible *material* (or *contingent*) *situations* of a certain concept of reality (see P. Mittelstaedt in this volume). The central notion of such a semantics is that of *valuation.* If a valuation assigns one of the two truth-values *true* and *false* to each sentence of the syntax, it is called a *bivaluation.* In normal propositional logic, the semantics of the language specifies a set of bivaluations corresponding to the possible material situations. This semantics can also be used for establishing a language for classical physics. Concerning a language for quantum physics it turns out that the classical concepts of the formal semantics of propositional logic cannot be taken over without additional *pragmatic* restrictions with respect to the semantics. For instance, a bivaluation is restricted to sets of those sentences which are pairwise *commensurable.* However this restriction itself does not come from the semantics but from the theory of quantum mechanical measurements. The commensurability and incommensurability conditions of quantum mechanical measurements confine the applicability of the classical semantics with respect to quantum mechanical sentences. In this way an object-language of quantum physics (which makes use of the classical concepts) possesses a semantical as well as a pragmatic component.

On the other hand, the dialogic approach to the language of quantum physics dispenses with the second component since it formulates the pragmatic restrictions as particular rules of argumentation and incorporates them into the dialogic semantics. As a main result we show in this contribution that a generalization of the classical semantics which makes use of the notion of bivaluation also dispenses with the pragmatic component to the object-language of quantum physics. The kind of extension of the semantics is immediately suggested by the dialogic semantics: The commensurability and incommensurability of propositions, which are formulated as propositions within the dialog-game, are considered to be propositions about particular properties of a quantum mechanical system; the language, in a generalized view, is regarded as a means for the description of *possible material processes*, which can be realized by sequences of proofs and disproofs of propositions within the runs of dialogs, and which can be considered to be possible material *situations* only under the condition of commensurability of the corresponding propositions. A set of generalized *valuations* of the language is established, corresponding to the possible material processes. Each valuation assigns one of the two truth-values to *some* sentences of the language, namely truth to those which are "proved" by the corresponding material process, and falsity to those which are "disproved" by the corresponding material process.

2. The Syntax of the Propositional Language

The syntax consists in a definition of the vocabulary and in a specification (a grammar) of how *sentences* are to be constructed from the vocabulary. The vocabulary comprises *elementary sentences a, b, . . .* , the *true* and *false sentences* $\bigvee$ and $\bigwedge$, the *commensurability* and *incommensurability symbols* $k(,)$ and $\bar{k}(,)$, and the *logical connectives* $\wedge, \vee, \rightarrow, \neg$ (which are already introduced by P. Mittelstaedt in this volume). In addition, we use the *sequential connectives* $\sqcap, \sqcup, \dashv$ as symbols for "and then", "or then", "if at first . . . then".

Thus we have the following syntax of the formal propositional language of quantum physics.

(2.1) **Definition**: A Q-propositional syntactic system (QPS) is a pair $\langle\langle S_a, C\rangle, S\rangle$, where

(a) S_a is the set of *atomic sentences*, which consists of *elementary sentences a, b, . . .* $\in S_e$, the *true* and *false* sentences $\bigvee$ and $\bigwedge$;

(b) C is the set $\{\wedge, \vee, \rightarrow, \neg, \sqcap, \sqcup, \dashv, k(,), k(,),), (\}$ of *logical* and *se-*

quential connectives, the *commensurability* and *incommensurability symbols* and auxiliary *parentheses*) , (.

(c) $\mathcal{S}$ is the set of *sequential sentences* ($\mathcal{S}$*-sentences*) $\mathsf{A}, \mathsf{B}, \mathsf{C}, \ldots$ which includes the set of *logical sentences* ($\mathcal{L}$*-sentences*) $A, B, C, \ldots$. $\mathcal{S}$ is recursively defined by:

(i) if $a \in \mathcal{S}_a$ then $a \in \mathcal{L}$;

(ii) if $A, B \in \mathcal{L}$ then $(A \wedge B), (A \vee B), (A \rightarrow B), \neg A, k(A, B), \bar{k}(A, B) \in \mathcal{L}$;

(iii) if $A \in \mathcal{L}$ then $A \in \mathcal{S}$;

(iv) if $\mathsf{A}, \mathsf{B} \in \mathcal{S}$ then $(\mathsf{A} \sqcap \mathsf{B}), (\mathsf{A} \sqcup \mathsf{B}), (\mathsf{A} \dashv \mathsf{B}), \neg \mathsf{A} \in \mathcal{S}$.

Elementary sentences $a \in \mathcal{S}_e$, and commensurability and incommensurability sentences $k(A, B), \bar{k}(A, B) \in \mathcal{L}$ are also called *material sentences.*

The above syntactic system contains a comprehensive set of connectives. Not all connectives are independent if the *value-definiteness* of material sentences is assumed. In this case an arbitrary pair of the 2-place logical connectives $\wedge, \vee, \rightarrow$ can be eliminated. Moreover, in QPS, all 2-place logical connectives and an arbitrary pair of sequential connectives can be eliminated. The connectives $\sqcap, \neg$ would be a minimal basis of connectives, for instance. In addition, $\bar{k}(A, B)$ may be replaced by $\neg\, k(A, B)$.

3. Hilbert Space Semantics for Material Sentences

The semantics comprises the possibilities of how the syntax can be *interpreted.* An *interpretation* of the syntax is a function which assigns to each sentence $\mathsf{A} \in \mathcal{S}$ a particular *proposition* $\tilde{\mathsf{A}}$. We say that the sentence A "expresses" the proposition $\tilde{\mathsf{A}}$. Since in the following we always refer to one possible interpretation, we use the same symbols to denote both the sentences and the corresponding propositions. In addition, the semantics comprises conditions with respect to the *truth* and *falsity* of the sentences relative to a given interpretation.

At first we consider a possible semantics of the set of *material* sentences. In a second step we consider the dialogic semantics of *compound* quantum mechanical sentences.

As an example for a semantics of material sentences we refer to the hilbert space formalism of standard quantum mechanics. Elementary quantum mechanical sentences are interpreted to be predictions of particular properties of a given quantum mechnical system. Such predictions, which can be proved by means of quantum mechanical measurements, are called *elementary propositions.* In von Neumann's theory a physical system S is represented by a hilbert space $\mathcal{H}(S)$. Each property $\hat{a}$ of the system is represented by a closed subspace M_a of $\mathcal{H}(S)$. Depending on the particular preparation, the system is characterized by a particular state. A system can in principle be prepared in a pure state which corresponds to a vector $|\varphi\rangle$ in the hilbert space.

If a system is given in a state $|\varphi\rangle$, a measurement of a property $\hat{a}$ is represented by a mapping which projects the vector $|\varphi\rangle$ orthogonally into the subspace M_a. This mapping defines the projection operator P_a on $\mathcal{H}$ with the range M_a. Hence, each property $\hat{a}$ of the system can equally be represented by the projection operator P_a.

In case P_a satisfies the relation $P_a|\varphi\rangle = |\varphi\rangle$, which is equivalent to $|\varphi\rangle \in M_a$, the property $\hat{a}$ will certainly be proved by a measurement. In case $P_a|\varphi\rangle = 0$ holds, which is equivalent to $|\varphi\rangle \in M_a^{\perp}$ where $M_a^{\perp}$ is the completely orthogonal subspace with respect to M_a, the property $\hat{a}$ will certainly be disproved by a measurement. In these two cases, the hilbert space formalism permits inferring the truth and falsity of an elementary sentence (which is associated with a prediction of a property of the system) already from the preparation of the system.

However, this is not possible for all elementary sentences. If $|\varphi\rangle \notin M_a$ and $|\varphi\rangle \notin M_a^{\perp}$, the theory of measurement does not specify the system to be in a pure state after the interaction with a measuring apparatus, but in a mixed state (e.g. see [2]). This mixed state comprises the two possibilities that, after "reading the scale", the system is in the state $|\psi\rangle$, which is the orthogonal projection of $|\varphi\rangle$ into M_a, or in the state $|\chi\rangle$, which is the orthogonal projection of $|\varphi\rangle$ into $M_a^{\perp}$.

The hilbert space formalism leads to the following proof and disproof conditions for elementary propositions. Let us assume that a system S is prepared in the state $|\varphi\rangle$, denoted by $S(\varphi)$. An elementary proposition states a certain property $\hat{a}$ of the system after its preparation in the state $|\varphi\rangle$. If, after the measuring process with respect to the property $\hat{a}$ (which includes reading the scale), the system is in the state $|\varphi'\rangle$, we have:

(3.1) **Definition**:
(a) The elementary sentence a is *true* in the system $S(\varphi)$ relative to a given interpretation

$\overset{\text{def.}}{\leftrightarrow}$ the corresponding elementary proposition $\hat{a}$ is proved by means of a measuring process M (denoted by $S(\varphi) \vdash_M a$)

$\leftrightarrow P_a|\varphi'\rangle = |\varphi'\rangle$;

(b) the elementary sentence $\hat{a}$ is *false* in the system $S(\varphi)$ relative to a given interpretation

$\overset{\text{def.}}{\leftrightarrow}$ the corresponding elementary proposition a is disproved by means of a measuring process M (denoted by $S(\varphi) \Vdash_M a$)

$\leftrightarrow P_a|\varphi'\rangle = 0$.

Commensurability and Incommensurability sentences $k(a, b)$ and $\bar{k}(a, b)$ are interpreted to be predictions of the commensurability and incommensurability of the elementary quantum mechanical propositions a and b (see P. Mittelstaedt in this volume). The hilbert space semantics is extended to commensurability sentences in the following way. Each commensurability sentence $k(a, b)$ is associated (via the interpretation) with the closed subspace $(M_a \cap M_b) \cup (M_a \cap M_b^\perp) \cup (M_a^\perp \cap M_b) \cup (M_a^\perp \cap M_b^\perp) =: M_{k(a,b)}$, where the operation $\cap$ denotes the intersection of subspaces and the operation $\cup$ denotes the span of subspaces. The correspondence of $k(a, b)$ to $M_{k(a,b)}$ may be motivated at this stage by the relation:

$$(3.2)\ |\varphi\rangle \in M_{k(a,b)} \curvearrowleft\!\!\curvearrowright [P_a, P_b]_- |\varphi\rangle = 0$$

which can easily be proved in the hilbert space formalism. The projection operator $P_{k(a,b)}$ onto $M_{k(a,b)}$ is given[1] by $P_{k(a,b)} := (s)\lim_{n\to\infty} (P_a \cdot P_b)^n + (s)\lim_{n\to\infty} (P_a \cdot (1 - P_b))^n + (s)\lim_{n\to\infty} ((1 - P_a) \cdot P_b)^n + (s)\lim_{n\to\infty} ((1 - P_a) \cdot (1 - P_b))^n$, such that

$$(3.3)\ P_{k(a,b)}|\varphi\rangle = |\varphi\rangle \curvearrowleft\!\!\curvearrowright |\varphi\rangle \in M_{k(a,b)}\ ;$$

$$(3.4)\ P_{k(a,b)}|\varphi\rangle = 0 \curvearrowleft\!\!\curvearrowright |\varphi\rangle \in M_{k(a,b)}^\perp .$$

Since we assume that $k(a, b)$ can be proved by quantum mechanical measurements, the above considerations of the theory of measurement can immediately be applied to commensurability propositions also:

(3.5) **Definition:**
(a) The commensurability sentence $k(a, b)$ is *true* in $S(\varphi)$ relative to a given interpretation

$$\overset{\text{def.}}{\curvearrowleft\!\!\curvearrowright}\ S(\varphi) \vdash_M k(a, b) \curvearrowleft\!\!\curvearrowright P_{k(a,b)}|\varphi'\rangle = |\varphi'\rangle\ ;$$

(b) the commensurability sentence $k(a, b)$ is *false* in $S(\varphi)$ relative to a given interpretation

$$\overset{\text{def.}}{\curvearrowleft\!\!\curvearrowright}\ S(\varphi) \vdash_M k(a, b) \curvearrowleft\!\!\curvearrowright P_{k(a,b)}|\varphi'\rangle = 0\ .$$

Analogous definitions can be given for the incommensurability sentence $\bar{k}(a, b)$.

4. Dialogic Semantics for Compound Sentences

The logical and sequential connectives of the propositional language of quantum physics are interpreted by the possibilities of argumentation within

1) The symbol (s)lim denotes the *strong convergence* of a sequence of operators.

quantum mechanical language games. In this way compound sentences are interpreted to be *compound propositions* about which a dialog-game can be performed. Since the theory of dialog-games also determines the truth conditions for logically and sequentially connected sentences, it provides the compound quantum mechanical sentences with a formal semantics.

It should be noted that the language game for quantum mechanical sentences has a structure which can be founded without reference to the particular interpretation of material quantum mechanical sentences [4, 5]. In these papers it is argued that a language for science has a certain discoverable skeleton which can adequately be reconstructed by means of a language game for science. This skeleton, as a *universal structure* of a scientific language. is a linguistic precondition of any scientific conception of the reality. In this sense it may be considered as a cognition which is a priori valid. It is a remarkable feature of the formal language of quantum physics that it is entirely established by this language game.

The language game can be formulated as a *dialog-game* (or a *two-person zero-sum game* in the terminology of game theory) $\underline{D}_m$, the rules of which are presented by P. Mittelstaedt in this volume. Since we also include the *sequential connectives* in the syntax of the language we have the following additional attack and defence scheme with respect to sequential propositions:

sequential connective	sequentially connected proposition	attacks	defences
sequential conjunction	$A \sqcap B$	1.1? 3.2?	2. A 4. B
sequential disjunction	$A \sqcup B$	1.?	2. A 3. B
sequential material implication	$A \dashv B$	1. A	2. B
negation	$\neg A$	A	

The succession of the attacks and defences is determined as is indicated by numbers in the above scheme.

The dialogic semantics, together with the hilbert space semantics, establishes a proof procedure for compound quantum mechanical propositions in a way which is illustrated by means of the following diagram:

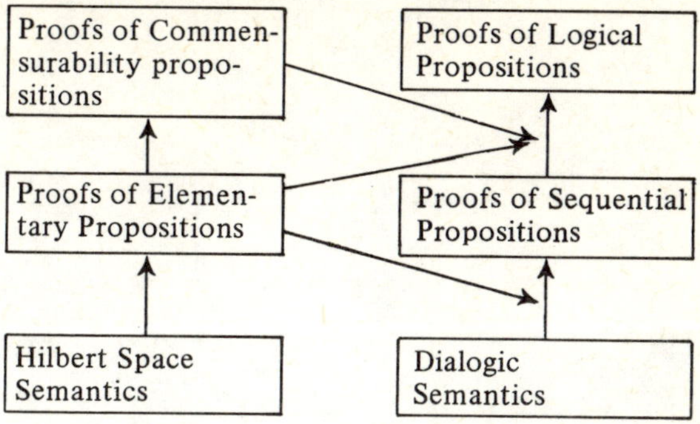

Truth and *falsity* of a compound sentence, relative to a given interpretation, are defined by:

(4.1) **Definition**:
(a) A sequential sentence A is *true* in the system $S(\varphi)$ relative to a given interpretation

$\overset{\text{def.}}{\Leftrightarrow}$ the proponent wins the dialog-game D_m about the corresponding proposition A (denoted by $S(\varphi) \vdash_{D_m} A$);

(b) a sequential sentence A is *false* in the system $S(\varphi)$ relative to a given interpretation

$\overset{\text{def.}}{\Leftrightarrow}$ the proponent loses the dialog-game D_m about the corresponding proposition A (denoted by $S(\varphi) \Vdash_{D_m} A$).

If a sentence is true (or false) relative to an interpretation, we also say that the corresponding proposition is true (or false).

5. The Admissible Valuations of the Language

The dialogic semantics is now used for specifying the set of admissible *valuations* of the language of quantum physics. As we shall show in the following, the dialogic proof procedure provides quantum mechanical sentences with a very natural valuation.

Each dialog consists in a succession of arguments which is determined by the dialog-rules up to the tests of material propositions. Due to the contingent proof results of material propositions, there are different possibilities of the continuation of the dialog. These possibilities may be represented by a game tree. An example is the game tree with respect to the proposition $a \wedge b$:

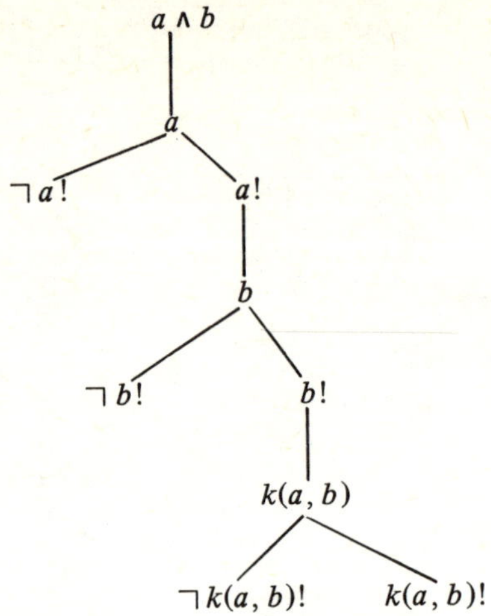

Disproofs of material propositions m are here denoted by $\neg m!$. Each individual dialog about $a \wedge b$ runs from the initial argument $a \wedge b$ to a final argument of the tree, depending on the contingent sequence of proof results of the material propositions. A possible dialog about $a \wedge b$ is for instance:

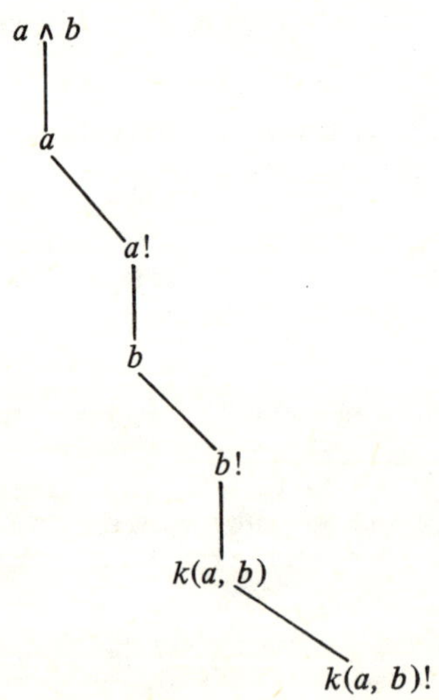

If the system is initially prepared in the state $|\varphi\rangle$, the state after the measuring process with respect to a is $|\varphi'\rangle$, the state after the measuring process with respect to b is $|\varphi''\rangle$, and the state after the measuring process with respect to $k(a, b)$ is $|\varphi'''\rangle$, we have for the above dialog:

$$P_a|\varphi'\rangle = |\varphi'\rangle, \quad P_b|\varphi''\rangle = |\varphi''\rangle, \quad P_{k(a,b)}|\varphi'''\rangle = |\varphi'''\rangle .$$

In the case of the above dialog, the sentence $a \wedge b$ is true. In the remaining three cases of possible dialogs about $a \wedge b$, the sentence $a \wedge b$ is false.

Within the framework of our semantics, material sentences are interpreted to be predictions of certain properties of a given physical system, which can be proved by means of operational testing procedures. We may consider the concept of *property* of a system as a primitive (undefined) concept of reality, which, in our example, is represented by means of the hilbert space formalism. We assume that, in case a material proposition m is disproved, the system has a certain property, denoted by $\neg \hat{m}$. Thus, the transition of a physical system, due to the proofs and disproofs of material propositions within the course of a dialog, can be characterized by means of a sequence of properties pertaining to the system. We call such a sequence a *possible material process.* Each individual dialog about a proposition is specified by a structural part, due to the connective structure of the proposition, and by a possible material process. For instance, the above dialog, in the course of which the proposition $a \wedge b$ is proved, is specified by the possible material process $\langle \hat{a}, \hat{b}, \hat{k}(a, b)\rangle$. A dialog, in the course of which the proposition $a \wedge b$ is disproved and therefore $\neg (a \wedge b)$ is proved, is specified by the process $\langle \hat{a}, \hat{b}, \neg \hat{k}(a, b)\rangle$ for instance.

Because of the possible incommensurability of quantum mechanical propositions, the truth-values of the material propositions within the run of a dialog cannot in general be determined by any knowledge of truth-values concerning the material propositions before the dialog. An entity which allows to predict the truth-values of all material propositions within the run of dialog is called a *material (contingent) situation.* In a classical language (e.g. the language of classical physics) the truth-values of all compound propositions are determined if a material situation is given. Therefore a classical language can be specified by a set of *bivaluations* corresponding to the material situations.

(5.1) **Definition:** A *bivaluation* is a function

$$v : S \longrightarrow \{T, F\}$$

such that for all sentences $A, B \in S$

(i) $v(A) \in \{T, F\}$;

(ii) $v(\neg A) = T \leftrightarrow v(A) = F$;

(iii) $v(A \wedge B) = T \leftrightarrow v(A) = T$ and $v(B) = T$.

A semantics for the object-language of quantum physics which makes use of the notion of valuation can now be established in the following way: We assume that, with respect to a quantum mechanical system S, the totality of all possible material processes forms a set $P(S)$. Each sentence A (relative to a certain interpretation of the elementary sentences by elementary propositions) is associated with a set $P(\mathsf{A})$ of all those possible material processes which specify all possible dialogs about the corresponding proposition A in the course of which A is proved. For instance we have $P(a \wedge b) = \{\langle \hat{a}, \hat{b}, \hat{k}(a, b)\rangle\}$ and $P(\neg (a \wedge b)) = \{\langle \neg \hat{a}\rangle, \langle \hat{a}, \neg \hat{b}\rangle, \langle \hat{a}, \hat{b}, \neg \hat{k}(a, b)\rangle\}$.

According to the definition of the sequential conjunction (in part 4), a possible material process p can uniquely be associated with an iterated sequential conjunction, denoted by $\sqcap p$, which consists of the material sentences, the corresponding propositions of which are proved, and the negations of the material sentences, the corresponding propositions of which are disproved. For instance, in this way the process $\langle \hat{a}, \hat{b}, \hat{k}(a, b)\rangle$ is associated with the sequential conjunction $(a \sqcap b) \sqcap k(a, b)$. In order to establish the set $P(\mathsf{A})$ of all possible material processes with respect to the sentence A, the dialog-game may be replaced by:

(5.2) The *algorithm* M:

(M 1) $\sqcap p \quad \Rightarrow p \qquad$ if $p \in P(S)$

(M 2) $(\mathsf{A}' \sqcap \neg (\mathsf{A} \sqcap \mathsf{B})) \sqcap \mathsf{B}' \Rightarrow (\mathsf{A}' \sqcap \neg \mathsf{A}) \sqcap \mathsf{B}',$, $((\mathsf{A}' \sqcap \mathsf{A}) \sqcap \neg \mathsf{B}) \sqcap \mathsf{B}'$;

together with the following possibilities of substitution ($\mathsf{A} = \mathsf{B}$ means that, within M, A may be substituted by B and vice versa):

(5.3)

(a) $(\mathsf{A} \sqcap \mathsf{B}) \sqcap \mathsf{C} = \mathsf{A} \sqcap (\mathsf{B} \sqcap \mathsf{C})$

(b) $\mathsf{V} \sqcap \mathsf{A} = \mathsf{A} = \mathsf{A} \sqcap \mathsf{V}$

(c) $\mathsf{A} \sqcup \mathsf{B} = \neg(\neg \mathsf{A} \sqcap \neg \mathsf{B})$

(d) $\mathsf{A} \dashv \mathsf{B} = \neg \mathsf{A} \sqcup \mathsf{B}$

(e) $\neg(\neg \mathsf{A}) = \mathsf{A}$

(f) $A \wedge B = (A \sqcap B) \sqcap k(A, B) \quad$ = commutations with respect to $\sqcap$

(g) $A \vee B = (A \sqcup B) \sqcup \neg k(A,B) \quad$ = commutations with respect to $\sqcup$

(h) $A \rightarrow B = A \dashv (B \sqcap k(A, B)) \quad = A \dashv (k(A, B) \sqcap B)$

(i) $k(A, B) = k(B, A) = k(\neg A, B)$

Starting with a sentence A and a given interpretation of the elementary subsentences of A by elementary propositions, an iterated application of the rule (M 2) and (5.3) leads to some sequence $\sqcap p, \sqcap p', \ldots$; which by means of the rule (M 1) establishes the set $P(\mathsf{A}) = \{p, p', \ldots\}$.

By means of the set $\mathcal{S}$ of sentences, the set $P(S)$ of all possible material processes with respect to a system S, and the sets $P(\mathsf{A})$ with $\mathsf{A} \in \mathcal{S}$, *truth* and *falsity* of sentences relative to a given interpretation are given by:

(5.4)

(a) *A* sentence A is *true in p* $\Leftrightarrow$ $p \in P(\mathsf{A})$;

(b) a sentence A is *false in p* $\Leftrightarrow$ $p \in P(\neg\, \mathsf{A})$.

A valuation of the sentences of the formal language of quantum physics can now be obtained by:

(5.5) **Definition:** A *valuation over P(S)* is the function

$$v : M \subseteq P(S) \times \mathcal{S} \rightarrow \{T, F\}$$

with $M := \{(p, \mathsf{A}) : \mathsf{A} \in \mathcal{S}, p \in P(\mathsf{A}) \cup P(\neg\, \mathsf{A})\}$

and:

(a) $v_p(\mathsf{A}) := v((p, \mathsf{A})) \doteq T \overset{\text{def.}}{\Leftrightarrow} \mathsf{A}$ is true in $p \in P(S)$;

(b) $v_p(\mathsf{A}) := v((p, \mathsf{A})) = F \overset{\text{def.}}{\Leftrightarrow} \mathsf{A}$ is false in $p \in P(S)$.

The function

$$v_p : \Delta_p \subseteq \mathcal{S} \longrightarrow \{T, F\}$$

with $\Delta_p := \{\mathsf{A} \in \mathcal{S} : p \in P(\mathsf{A}) \cup P(\neg\, \mathsf{A})\}$

is called an *admissible valuation on* $\mathcal{S}$. The set of admissible valuations (QVL_i) on $\mathcal{S}$ relative to a given interpretation *i* is the set $\{v_p : p \in P(S)\}$.

This establishes the language of quantum physics relative to a given interpretation *i* of the syntax, QL_i, as a pair $\langle QPS, QVL_i \rangle$, where *QPS* in the *Q-propositional syntactic system* (2.1) and QVL_i is the set of admissible valuations on $\mathcal{S}$. The concept of *satisfaction* with respect to a sentence can be defined as usual (e.g. see [7]): A sentence A is said to be *satisfied* by the admissible valuation v_p if and only if $v_p(\mathsf{A}) = T$. A is said to be *satisfiable* if and only if there exists a possible material process $p \in P(S)$ such that $v_p(\mathsf{A}) = T$. The concepts of *refutation* and *refutability* can be defined analogously. We say that a sentence A is *always true* relative to an interpretation *i*, if and only if $P(\neg\, \mathsf{A}) = \emptyset$, i.e. $v_p(\mathsf{A}) = T$ for all $p \in P(\mathsf{A}) \cup P(\neg\, \mathsf{A})$ (always *false* or *not satisfiable* analogously). This semantic concept gives rise to the modality of *necessity*.

In his contribution to this volume, P. Mittelstaedt considers the modalities as meta-linguistic concepts using an extension of the dialog-game semantics. However, it is also possible to introduce the modalities into the object-language without extending the semantics. This can be done by adding the sign Δ to the set *C* of the syntax (see (2.1)) and by extending the set $\mathcal{S}$ of sentences to:

(v) if $\mathsf{A} \in \mathcal{S}$ then $\Delta \mathsf{A} \in \mathcal{S}$.

Since truth is defined relative to a possible material process p, and, throughout in our above considerations, relative to a given interpretation of the syntax, we stipulate:

A sentence ΔA is *true in p*

$\overset{\text{def.}}{\curvearrowleft\curvearrowright}$ A is true in p and in all $p' \in P(A) \cup P(\neg A)$.

In order to extend the admissible valuations to sentences of the kind ΔA, we use $P(\Delta A) \rightleftharpoons P(A)$, and then we take over the definition (5.5) for the extended set S.

The most general representation of the language of quantum physics takes into account all possible interpretations of the syntax by means of elementary propositions. Therefore we define the *formal language* as a pair $\langle QPS, \langle I, QVL \rangle\rangle$, where I is the set of *admissible interpretations* (i.e. all possible assignments of elementary propositions to elementary sentences), and QVL is the set of sets of admissible valuations corresponding to the interpretations. With respect to the formal language a sentence A is said to be *satisfiable* if and only if there exists an interpretation $i \in I$ such that A is satisfiable relative to i. We say that A is *formally true* if and only if the sentence A is always true *for all admissible interpretations*. The concept of formal truth defined here can be considered to be a generalization of the concept of *validity* in classical propositional logic. In this logic, a sentence A is defined to be *valid* if and only if A is satisfied by *all* admissible valuations relative to all admissible interpretations of the syntax. However, this concept is only applicable in case a sentence belongs to the domain of each admissible valuation. As definition (5.5) shows, this condition is not satisfied by the language of quantum physics.

When developing the language of quantum physics on the basis of the dialogic semantics we did not take into account (for the sake of simplicity) that certain material processes are *identical* in the sense that they cannot be distinguished within the dialogic semantics. Thus, equivalence classes of possible material processes instead of possible material processes should be used for establishing a valuation of the language. (cf. [6]).

6. Concluding Remarks

The formal language of quantum physics, considered in this contribution, allows to specify those sentences which are true according to the particular concepts of truth. If a sentence A is *formally true* we write $\Vdash A$. In order to specify the sentences which are formally true, one usually proceeds to a *logical system* which is adequate to the language. A logical system comprises the syntax of the language and a set of theorems $\emptyset \vdash A$, where $\emptyset$ is the empty set and A is a sentence. Adequateness of the logical system to the formal language means that $\Vdash A \leftrightarrow \emptyset \vdash A$ is satisfied, i.e. that the logical system is *complete* and *sound* with respect to the formal language.

A logical system which is adequate to the formal language of quantum physics has been established in [6]. The set of theorems of this system is constructively given by means of a calculus which allows to deduce the theorems of the logical system. An algebraic representation of the logic can now be obtained by generating the *Lindenbaum-Tarski algebra* of the calculus. It turns out, as is shown in [6], that this algebra is a *Baer*-semi-group* with respect to the operation of sequential conjunction and with respect to two additional derived unary operations. With respect to the logical operations the Lindenbaum-Tarski algebra possesses a *free ortho-modular lattice* as a substructure. Baer*-semigroups and orthomodular lattices have been investigated in connection with their realization by operator algebras on a hilbert space. It is well known that the algebra of projection operators on a hilbert space is an orthomodular lattice. In the framework of the *operational approach* to quantum mechanics [3], certain entities, called *operations*, form a Baer*-semigroup and are realized by products of projection operators on a hilbert space. The *logical approach* to quantum physics, which starts with an analysis of the object-language of quantum physics, leads to the following realization of sentences: logically connected sentences are realized by projection operators on a hilbert space; sequentially connected sentences are realized by a subalgebra of the algebra of continuous linear operators on a hilbert space which is generated in a particular way by the set of projection operators and the operator multiplication and addition (cf. [6]).

References:

[1] Lorenzen, P., and Lorenz, K. (1978), *Dialogische Logik*, Wissenschaftliche Buchgesellschaft, Darmstadt

[2] Mittelstaedt, P. (1976), *Philosophical Problems of Modern Physics*, D. Reidel Publ. Co., Dordrecht, Holland, Chapt. III.

[3] Pool, J. C. T., (1968), 'Baer*-Semigroups and the Logic of Quantum Mechanics', *Communications on Mathematical Physics* 9, 118–141; reprinted in C. A. Hooker (ed.), (1975), *The Logico-Algebraic Approach to Quantum Mechanics*, D. Reidel Publ. Co., Dordrecht, Holland, 365–394

[4] Stachow, E.-W. (1976) 'Completeness of Quantum Logic, *Journal of Philosophical Logic 5*, 237–280. Reprinted in C. A. Hooker (ed.), (1979), *Physical Theory as Logico-Operational Structure*, D. Reidel, Dordrecht, Holland, 203–243.

[5] Stachow, E.-W. (1979) 'On a Game-Theoretical Approach to a Scientific Language' *PSA 1978*, Volume 2 (1979).

[6] Stachow, E.-W. (1980) 'On the Logical Foundation of Quantum Mechanics', *International Journal of Theoretical Physics*, Vol. 19, No. 4, 251–304.

[7] van Fraassen, B. C. (1971), *Formal Semantics and Logic*. The Mac-Millan Co., New York.

A Unified Concept of Evolution in Quantum Mechanics

by

C. Piron

It is well known that there are two kinds of evolution in Quantum Mechanics. This was emphasized by J. von Neumann in his book. There is the unitary one defined by the Schrödinger equation and describing the evolution of the isolated system and the so called reduction of the wave packet, defined by an orthogonal projection, and describing the more or less instantaneous effect of the measurement process. We have reconsidered the basis of the Schrödinger equation recently and we have discovered, as a by-product a unified description of these two appearently essentially different evolution processes. In order to explain the motivations of a reconsideration of the Schrödinger evolution, let us first consider the case of a classical mechanical system like a particle. For such a system one poses in general the following problem: Determine the final state given the initial state and the exterior field which acts on the particle. This is not a well defined problem, given the initial state it is possible to calculate the corresponding orbit in the phase space but not the final state. To answer the question completely we must say something about the final state for example the final time $t = t_1$, the final first position coordinate $q_1 = x_1$, or the final value of some other variable. We will call such a variable the control variable since it is clear that in the corresponding experimental situation we must continuously look at this variable during all the evolution Given a particle of mass M in an exterior field $(\vec{A}(\vec{q}, t), V(\vec{q}, t))$, to describe the evolution in a way independent of the given control variable we can proceed as follows:

In the phase space $\Gamma \equiv \{(\vec{p}, E, \vec{q}, t)\}$ we consider the restriction Ω_K of the 2-form

$$d\vec{p} \wedge d\vec{q} - dE \wedge dt$$

on the manifold Γ_K defined by

$$K \equiv \frac{1}{2M}(\vec{p} - \vec{A}(\vec{q}, t))^2 + V(q, t) - E = 0$$

in the Galilean case and by

$$K \equiv \frac{1}{2M}[(\vec{p} - \vec{A}(\vec{q}, t))^2 - \frac{1}{c^2}(E - V(\vec{q}, t))^2 + M^2c^2] = 0$$

in the relativistic case. The evolution is defined by the tangent vector field $\eta \in T\Gamma_K$ such that

$$\Omega_K(\eta, \xi) = 0 \quad \forall \quad \xi \in T\Gamma_K$$

If we choose t as the control variable, we write E as a function of the other variables:

$$E = H(\vec{p}, \vec{q}, t) \equiv \frac{1}{2M} (\vec{p} - \vec{A}(\vec{q}, t))^2 + V(\vec{q}, t)$$

or

$$E = H(\vec{p}, \vec{q}, t) \equiv c\, [(\vec{p} - \vec{A}(\vec{q}, t))^2 + M^2c^2]^{1/2} + V(\vec{q}, t)$$

and we find for the canonical equations:

$$\frac{dp_i}{dt} = -\frac{\partial H}{\partial q^i}, \quad \frac{dE}{dt} = \frac{\partial H}{\partial t}, \quad \frac{dq^i}{dt} = \frac{\partial H}{\partial p_i}.$$

But if we choose q_1 as the control variable x we must write p_1 as function of the other variables

$$p_1 = \pi(p_2, p_3, E, q, t) \equiv [2M(E - V(\vec{q}, t)) - (p_2 - A_2(\vec{q}, t))^2 - (p_3 - A_3(\vec{q}, t))^2]^{1/2} + A_1(\vec{q}, t)$$

or

$$p_1 = \pi\, (p_2, p_3, E, \vec{q}, t) \equiv [\frac{1}{c^2} (E - V(\vec{q}, t))^2 - M^2c^2 - (p_2 - A_2(\vec{q}, t))^2 - (p_3 - A_3(\vec{q}, t))^2]^{1/2} + A_1(\vec{q}, t)$$

and we find the canonical equations:

$$\frac{dp_2}{dx} = \frac{\partial \pi}{\partial q^2}, \quad \frac{dp_3}{dx} = \frac{\partial \pi}{\partial q^3}, \quad \frac{dE}{dx} = -\frac{\partial \pi}{\partial t},$$

$$\frac{dq2}{dx} = -\frac{\partial \pi}{\partial p_2}, \quad \frac{dq^2}{dx} = -\frac{\partial \pi}{\partial p_3}, \quad \frac{dt}{dx} = \frac{\partial \pi}{\partial E}.$$

Let us consider the quantum case. We want first to make the two following remarks:

i) From the Bohr complementarity principle, if we are continuously measuring the observable corresponding to the control variable we cannot measure at the same time the complementary observable. The choice of the control variable is then not arbitrary but must correspond to the given experimental situation.

ii) The Hilbert spaces which describe the system describe in fact the possible results of experiments that we can perform or not perform on the system. So these do not describe what is given with the system like the observable corresponding to the control variable and the choice of the Hilbert space structure depends on the control variable.

In this spirit if t is the control variable a particle of spin 0 is described by a family of Hilbert spaces $\mathcal{H}_t$ all isomorphic to the space $L^2(\mathbb{R}^3, dq^1, dq^2, dq^3)$. On the contrary, if x is the control variable the same particle is described by a family of Hilbert spaces $\mathcal{H}_x$ all isomorphic to the space $L^2(\mathbb{R}^3, dq^2, dq^3, dt)$.

In the first case the usual observables are:

$$\begin{aligned} \underline{q}^i \varphi_t(\vec{q}) &= q^i \varphi_t(\vec{q}) \\ \underline{t}\, \varphi_t(\vec{q}) &= t\, \varphi_t(\vec{q}) \\ \underline{p}_i \varphi_t(\vec{q}) &= -i\hbar\partial_i \varphi_t(\vec{q}) \end{aligned}$$

and the evolution is defined by the Schrödinger equation:

$$i\hbar\partial_t \varphi_t(\vec{q}) = H(\underline{\vec{p}}, \underline{\vec{q}}, t)\, \varphi_t(\vec{q}) \qquad \text{(A)}$$

where $H(\vec{p}, \vec{q}, t)$ is given by the corresponding function defined for the classical case. In the second case the usual observables are:

$$\begin{aligned} \underline{q}^1 \varphi_x(q^2, q^3, t) &= x\, \varphi_x(q^2, q^3, t) \\ \underline{q}^2 \varphi_x(q^2, q^3, t) &= q^2 \varphi_x(q^2, q^3, t) \\ \underline{q}^3 \varphi_x(q^2, q^3, t) &= q^3 \varphi_x(q^2, q^3, t) \\ \underline{t}\, \varphi_x(q^2, q^3, t) &= t\, \varphi_x(q^2, q^3, t) \\ \underline{p}_2 \varphi_x(q^2, q^3, t) &= -it\,\partial_2 \varphi_x(q^2, q^3, t) \\ \underline{p}_3 \varphi_x(q^2, q^3, t) &= -i\hbar\partial_3 \varphi_x(q^2, q^3, t) \\ \underline{E}\, \varphi_x(q^2, q^3, t) &= i\hbar\partial_t \varphi_x(q^2, q^3, t) \end{aligned}$$

and the evolution is defined by the generalized Schrödinger equation:

$$-i\hbar\partial_x \varphi_x(q^2, q^3, t) = \pi(\underline{p}_2, \underline{p}_3, x; \underline{q}^2, \underline{q}^3, \underline{t})\, \varphi_x(q^2, q^3, t) \qquad \text{(B)}$$

where $\pi(p_2, p_3, x, q^2, q^3, t)$ is given by the corresponding function defined in the classical case.

Problem: How to characterize these two evolution equations (A) and (B) in an intrinsic way? This can be done according to the following remark. Let us consider the Hilbert space $L^2(\mathbb{R}^4, dq^1, dq^2, dq^3, dt)$ which can be obtained as the direct integral of the $\mathcal{H}_t$ as well as of the $\mathcal{H}_x$. If $\varphi_t(\vec{q})$ is solution of the evolution equation (A) then $\varphi(\vec{q}, t) = \varphi_t(\vec{q})$ is a generalized eigenvector of $K(\underline{\vec{p}}, \underline{E}, \underline{\vec{q}}, \underline{t})$ for the eigenvalue 0. For the same reason if $\varphi_x(q^2, q^3, t)$ is solution of the evolution equation (B) then $\varphi(\vec{q}, t) =$

$\varphi_x(q^2, q^3, t)$ is also a generalized eigenvector of the same $K(\underline{\vec{p}}, \underline{E}, \underline{\vec{q}}, \underline{t})$ for the same eigenvalue. This shows us that the Schrödinger evolution can be characterized by an eigenvector equation as a measurement process. Since there is in fact no eigenvector we must define a dense set $\mathcal{D}$ such that its dual, which is larger than $L^2(\mathbb{R}^4, dq^1 dq^2 dq^3 dt)$, contains the generalized eigenvector. To each choice of the control variable x there corresponds a well defined dense set $\mathcal{D}_x$, a decomposition into a family of Hilbert spaces and finally a reducible system of propositions.

As a conclusion, let us discuss an example of the application of the evolution Schrödinger equation (B). Let us consider a free quantum particle in the two dimensional space-time (in one dimension, in the old language). Given the initial state we want to know at what time a counter $x = x_1$ detects the particle. The counter is left on during all the experiment. Since in this case we measure t at the point x, and not x at time t, as usually, we have to take x as the control variable. The initial state will be given by $\varphi_{x_0}(t)$ and the Schrödinger evolution equation will be:

$$-i\hbar\,\partial_x\,\varphi_x(t) = (2M\underline{E})^{1/2}\,\varphi_x(t)$$

The solution is easily obtained

$$\varphi_{x_1}(t) = e^{i\hbar^{-1}(2M\underline{E})^{1/2}(x_1 - x_0)}\,\varphi_{x_0}(t)$$

The expectation value of the observable $\underline{t}$ at the point x_1, is given by

$$\begin{aligned}\langle \underline{t}\rangle_{x_1} &= \int dt\,\varphi^*_{x_1}(t)\,t\,\varphi_{x_1}(t)\\ &= \langle \underline{t}\rangle_{x_0} + \langle \frac{M}{(2M\underline{E})^{1/2}}\rangle_{x_0}(x_1 - x_0)\end{aligned}$$

since we have the commutation relation:

$$\begin{aligned}&\underline{t}\,e^{i\hbar^{-1}(2M\underline{E})^{1/2}(x_1 - x_0)}\\ &= e^{i\hbar^{-1}(2M\underline{E})^{1/2}(x_1 - x_0)}\left(\underline{t} + \frac{M}{(2M\underline{E})^{1/2}}(x_1 - x_0)\right)\end{aligned}$$

We can calculate the fluctuations of $\underline{t}$ also:

$$\begin{aligned}\langle \underline{t}^2\rangle_{x_1} - \langle \underline{t}\rangle^2_{x_1} &= \langle \underline{t}^2\rangle_{x_0} - \langle \underline{t}\rangle^2_{x_0}\\ &+ \left(\langle [t, \frac{M}{(2M\underline{E})^{1/2}}]_+\rangle_{x_0} - 2\,\langle \underline{t}\rangle_{x_0}\,\langle \frac{M}{(2M\underline{E})^{1/2}}\rangle_{x_0}\right)(x_1 - x_0)\\ &+ \left(\langle \frac{M^2}{2M\underline{E}}\rangle_{x_0} - \langle \frac{M}{(2M\underline{E})^{1/2}}\rangle^2_{x_0}\right)(x_1 - x_0)^2\end{aligned}$$

Conditional Expectations and Duals of Instruments

by

K.-E. Hellwig

Usually, quantal conditional expectations are introduced by the requirement of some algebraic and analytic properties known to hold in the classical theory. The resulting objects do not depend on a given state which is in contrast to the classical theory. Hence, the question arises whether these objects are the most appropriate analogues of the classical ones. – We proposed an alternative setting by going back to two more fundamental axioms for classical conditional expectations, transducing them into the language of operational approach to quantum theory and proposing generalized conditional expectations which seem to be closer analogues to the classical ones [H. Cycon, K.-E. Hellwig (1977)]. – In the present contribution a Hilbert space formulation is given and further results on the formulation of quantum stochastic processes based on generalized conditional expectations are presented. In particular, sufficient conditions on a one-parameter family of self adjoint operators such that a Chapman-Kolmogorov equation holds true are derived.

Conditional expectations have a basic meaning in the theory of classical stochastic processes which can be understood as a theory of successive measurements with random outcome on a physical system. This view is based on the classical nature of these measurements. Successive measurements also can be performed on quantum systems and from the peculiar nature of such measurements the problem arises whether there is a theory of quantum stochastic processes and how such a theory can be formulated. Experimental evidence for such theory can possibly be found in the physics of optical coherence, where up to now classical stochastic theory is applied to photon counting experiments.

Several attempts have been made to formulate quantum stochastic processes. In *C*-formulation* of quantum theory an a posteriori characterization of classical conditional expectation [Shu-Teh Chen Moy (1954)] has been used as the starting point [H. Umegaki (1954, 1956, 1959); M. Nakamura, T. Turumaru (1954); M. Nakamura, M. Takesaki (1960)]. Other attempts are based on the quantum theory of successive measurements (E. B. Davies, J. T. Lewis (1970); C. M. Edwards (1970, 1971); E. B. Davies (1976); M. D. Srinivas (1975, 1979),}. Now, quantal measurements are represented by certain linear operators, called operations, of the space of real trace class operators of some Hilbert space $\mathcal{H}$ which, in a well known sense, is the dual of the space of bounded hermitean operators of $\mathcal{H}$. Since operations act on density operators and conditional expectations shall act on observables, it is suggesting to dualize the action of that operation, by which the measurement is described. Such an adjoint operator of an operation is taken to be the corresponding conditional expectation. This ansatz has been worked out in [E. B. Davies, J. T. Lewis (1970); E. B. Davies (1976)].

Although the latter ansatz is tempting, it cannot be a strict generalization of the classical concept. Firstly, one observes that it does not depend on an initial density operator whereas the classical conditional expectation does depend on an initial probability measure. Secondly, if one dualizes the action of classical conditional expectation one obtains mappings of probability measures into themselves, but this mapping does not describe classical ideal measurements. We will come back to this point in sections 1 and 2.

Motivated by this gap, we gave an alternative formulation of conditional expectations of quantum theory in a previous paper [H. Cycon, K.-E. Hellwig (1977)] which is formulated in the language of E. B. Davies and J. T. Lewis (1970). To that end we went back to the fundamental definition of conditional expectations in classical probability theory as can be found in H. Bauer (1968). We could prove the existence of our generalized conditional expectation under certain assumptions but were not able to prove uniqueness.

In the present paper we confine ourselves to Hilbert space formulations and attempt a further step towards the formulation of quantum stochastic processes on the basis of our generalized conditional expectation. More precisely, we try to formulate conditional independence or, what comes to the same at least in the classical case, we try to find an analogue of the Chapman-Kolmogorov equation. Since we do not explore all of the freedom which is present in our model but restrict ourselves on the simplest case, the result is that such process must be classical when Chapman-Kolmogorov's equation holds true. But by the argument just given, nothing can be stated about general existence or non-existence of non-classical Markov processes. In section 1 we review the basic notions of the classical theory. In section 2, the quantum case, we have chosen an argumentation which runs along the lines of that given in section 1.

1. Conditional expectations and Markov processes in classical probability theory

In this section we give a brief review of the main concepts which we intend to generalize to the quantum case. Let be given a probability space $(\Omega, \mathcal{A}, \mu)$, Ω being a set, $\mathcal{A}$ being a σ-algebra of subsets of Ω, μ being a probability measure on $\mathcal{A}$. Further let be given two measurable mappings X and Y of Ω into some topological spaces $\mathcal{X}$ and $\mathcal{Y}$, respectively. We assume them to be locally compact and of second countability for definiteness of the following constructions, but we do not enter into mathematical details. Let $\mathcal{B}(\mathcal{X})$ and $\mathcal{B}(\mathcal{Y})$ denote the Borel sets of $\mathcal{X}$ and $\mathcal{Y}$, respectively. The elements of $\mathcal{B}(\mathcal{X})$ and $\mathcal{B}(\mathcal{Y})$ are interpreted as possible outcomes of measurements to be performed successively, e.g. that corresponding to X after that corresponding to Y. Instead of asking for the probability of some $\xi \in \mathcal{B}(\mathcal{X})$ conditioned by previous occurrence of $\eta \in \mathcal{B}(\mathcal{Y})$, we ask more generally for the expectation value of some bounded random variable f on $\mathcal{X}$ conditioned by previous occurrence of $\eta \in \mathcal{B}(\mathcal{Y})$. The point is to find a random variable $E^Y(f)$ on $\mathcal{Y}$ such that the conditioned expectation value of f for arbitrary $\eta \in \mathcal{B}(\mathcal{Y})$ is given by

$$\int_{Y^*(\eta)} E^Y(f) \circ Y \, d\mu ,$$

where $Y^*(\eta) := \{\omega \in \Omega \mid Y(\omega) \in \eta\}$. Clearly, we have presupposed that the observation of η to occur does not alter the probability measure μ. If $E^Y(f)$ exists for a measurable function f on the value space $\mathcal{X}$ of some observable X to be measured later, probability statements on subensembles arising from selection due to a previous measurement of Y can be expressed by the expectation value of some random variable $E^Y(f)$ on the probability space $(\mathcal{Y}, \mathcal{B}(\mathcal{Y}), \mu^Y)$, where $\mu^Y := \mu \circ Y^*$ such that

$$\int_{Y^*(\eta)} E^Y(f) \circ Y \, d\mu = \int_{\eta} E^Y(f) \, d\mu^Y$$

This seems to be very natural since $(\mathcal{Y}, \mathcal{B}(\mathcal{Y}), \mu^Y)$ is the very probability space for observation of outcomes of Y. The situation can be made graphic by the diagram

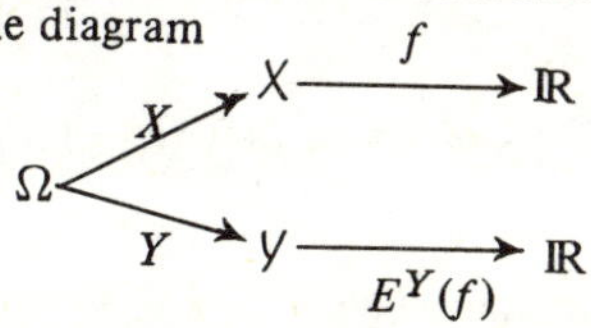

(1)

This background philosophy motivates the two postulates {H. Bauer (1968)}

$$E^Y(f) \text{ is a random variable on } \mathcal{Y} , \tag{I$_c$}$$

$$\int_{Y^*(\eta)} f \circ X \, d\mu = \int_{Y^*(\eta)} E^Y(f) \circ Y \, d\mu . \tag{II$_c$}$$

The random variable $E^Y(f)$ is easily constructed by use of the Radon-Nikodym theorem: Assuming $f \circ X$ to be integrable with respect to μ (this is more general than boundedness of f assumed above), we find $\int_{Y^*(.)} f \circ X \, d\mu$ to be a measure on $\mathcal{B}(\mathcal{Y})$ which is subordinated to $\mu^Y(.)$.

So $E^Y(f)$ can be taken to be a version of the Radon-Nikodym density of $\int_{Y^*(.)} f \circ X \, d\mu$, which is unique up to equivalence with respect to μ^Y. So we have the diagram

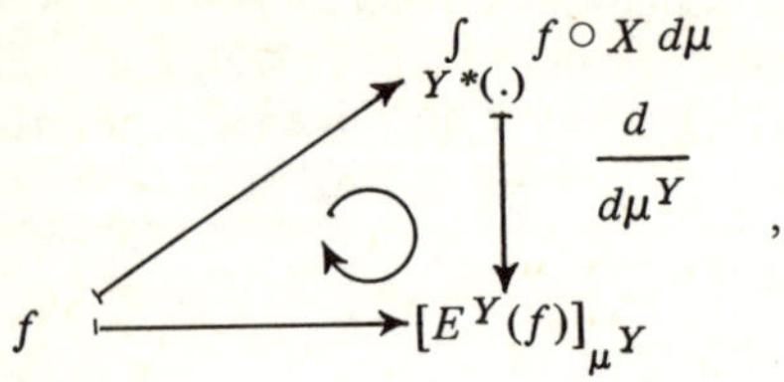

,

where $[E^Y(f)]_{\mu^Y}$ denotes the class of equivalent functions from which $E^Y(f)$ is a version. In order to keep notations easy, we omit the brackets in the following and keep in mind that some objects are only defined up to equivalence.

Assuming X locally compact and of second countability one can show that conditional expectations can be written as

$$E^Y(f) = \int_{\mathsf{X}} f \, \mu^{(X/Y)}(., dx) \,, \tag{3}$$

where

$$\mu^{(X/Y)} : \mathsf{B}(\mathsf{X}) \longrightarrow \mathsf{L}^{\infty}_{\mu^Y}(\mathsf{Y})$$

is a Markov kernel and called a conditional distribution of X with respect to Y. The number $\mu^{(X/Y)}(y, \xi)$, where $y \in \mathsf{Y}$, $\xi \in \mathsf{B}(\mathsf{X})$, is called the transition probability from y into ξ. By construction we have for the joint probability of $X^*(\xi)$ and $Y^*(\eta)$

$$\mu(X^*(\xi) \cap Y^*(\eta)) = \int_{Y^*(\eta)} \chi_{X^*(\xi)} d\mu = \int_{\eta} E^Y(\chi_\xi) \, d\mu^Y = \int_{\eta} \mu^{(X/Y)}(y, \xi) d\mu^Y \,, \tag{4}$$

where χ denotes the characteristic function of the set by which it is indexed.

Let us remark something about the operational view and the duality of instruments and expectations. Joint probabilities arise from selection procedures in connection with successive measurements, and such procedures can be described by instruments in the sense of E. B. Davies. The selection procedure which we have used above in our background philosophy is a classical ideal one. It is described by the mapping

$$(\eta, \upsilon) \longmapsto \upsilon(Y^*(\eta) \cap (.)) \,,$$

$\eta \in \mathsf{B}(\mathsf{Y})$, υ being a probability measure on A. We consider the conditional expectation of $f \circ X$ with respect to Y on the measurable space (Ω, A) which we define by

$$\mathbb{E}^Y(\eta, f \circ X) := (E^Y(f) \cdot \chi_\eta) \circ Y \,. \tag{5}$$

By interchanging of the order of integration we find

$$\int_{\Omega} \mathbb{E}^{Y}(\eta, f \circ X)\, d\upsilon = \int_{\eta} E^{Y}(f)\, d\upsilon^{Y} =$$

$$= \int_{\eta} \left(\int_{X} f(x)\, \mu^{(X/Y)}(y, dx)\right) d\upsilon^{Y} = \int_{X} f(x) \left(\int_{\eta} \mu^{(X/Y)}(y, dx)\, d\upsilon^{Y}\right).$$

Let now M_A, $M_{X^*[B(X)]}$ denote the convex sets of probability measures on the σ-algebras by which they are indexed and define

$$J : B(Y) \times M_A \to M_{X^*[B(X)]},$$

by

$$J(\eta, \upsilon)\,(X^*(\xi)) := \int_{\eta} \mu^{(X/Y)}(y, \xi)\, d\upsilon^{Y}.$$

Although the image of J is not M_A but $M_{X^*[B(X)]}$, we will call it by abuse of language an instrument since all other properties of instruments are fulfilled. In any case it should be noticed, that in general $J(\eta, \upsilon)$ is unequal to the ideal instrument, i.e.

$$J(\eta, \upsilon)\,(X^*(\xi)) \neq \upsilon(Y^*(\eta) \cap X^*(\xi)).$$

The instrument J is found dualizing the action of conditional expectations. Conversely, quantal conditional expectations are sought for in the literature by adoption of the operational view and dualization of instruments. Our argument shows here that this suggesting procedure leads to an alternative meaning of conditional expectations when applied to the classical case. We will see in section 2 that this argument holds equally true for the quantal case.

Now conditional expectations are a useful tool when Markov processes shall be formulated. To that end, another basic construct is conditional independence. Let be given a further measurable function Z on (Ω, A) with value space Z which also has properties like X and Y. Assume Z to be measured prior to Y. The situation can be made graphic by the following diagram:

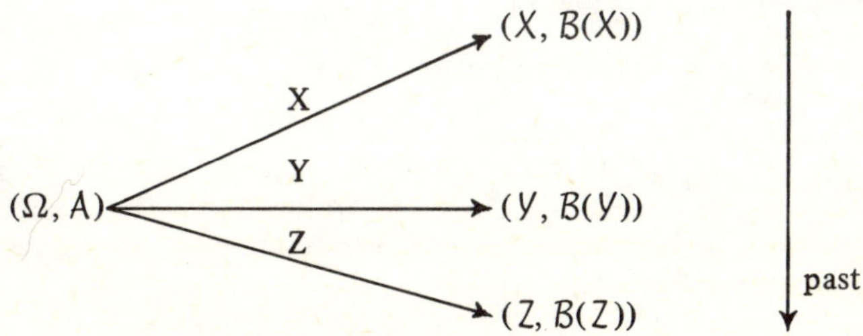

Generalization of the formula

$$\mu(X^*(\xi) \cap Y^*(\eta)) = \int_{\eta} \mu^{(X/Y)}(y, \xi)\, d\mu^Y \tag{6}$$

to the case of three subsequent measurements leads to an especially simple result, when conditional independence of X from Z with respect to Y is postulated, or, equivalently stated, when the Chapman-Kolmogorov equation

$$\mu^{(X/Z)}(.\,, \xi) = \int_{Y} \mu^{(X/Y)}(y, \xi)\, \mu^{(Y/Z)}(., dy) \tag{7}$$

holds true. In that case

$$\mu(X^*(\xi) \cap Y^*(\eta) \cap Z^*(\zeta)) = \int_{\zeta} \int_{\eta} \int_{\xi} \mu^{(X/Y)}(y, dx)\, \mu^{(Y/Z)}(z, dy)\, d\mu^z . \tag{8}$$

So far, we have considered some basic points of the construction of classical Markov processes. Roughly speaking, classical Markov processes arise by generalization from three observables $\{X, Y, Z\}$ to some one-parameter family $\{X_\tau\}_{\tau \in \mathbb{R}_0^+}$ of observables such that, whenever $r < s < t$ then equation (7) holds true for $Z := X_r$, $Y := X_s$, and $X := X_t$. Clearly, some regularity assumptions for kernels to exist have to be added, which is nontrivial but more or less technical. In section 2, we will try to give a quantum mechanical formulation of the simplified model with three observables.

We conclude this section by rewriting the formulae (6), (7), and (8) for joint probability in terms of the $\mathbb{E}^{\cdot}(.\,,.)$, which is appropriate for developments in the quantum case.

From (5) and (4) we find for (6)

$$\mu(X^*(\xi) \cap Y^*(\eta)) = \int \mathbb{E}^Y(\eta, \chi_\xi \circ X)\, d\mu . \tag{9}$$

By (3) and (5) we rewrite (7) as

$$\mathbb{E}^Z(\zeta, \chi_\xi \circ X) = \mathbb{E}^Z(\zeta, \mathbb{E}^Y(Y, \chi_\xi \circ X)) , \tag{10}$$

since the right hand side is equal to

$$\mathbb{E}^Z(\zeta, (\chi_Y \cdot \int_X \chi_\xi(x)\, \mu^{(X/Y)}(., dx)) \circ Y) =$$

$$= (\chi_\zeta \cdot \int_Y \int_X \chi_\xi(x)\, \mu^{(X/Y)}(y, dx) \mu^{(Y/Z)}(., dy)) \circ Z) =$$

$$(\chi_\zeta \cdot \int_X \chi_\xi(x)\, \mu^{(X/Z)}(., dx)) \circ Z$$

which is just the left hand side of (10). For (8) we can write

$$\mu(X^*(\xi) \cap Y^*(\eta) \cap Z^*(\zeta)) = \int \mathbb{E}^Z(\zeta, \mathbb{E}^Y(\eta, \chi_{X^*(\xi)}))\, d\mu\,. \tag{11}$$

2. Conditional expectations and Markov processes in quantum theory

The basic set Ω of classical probability theory does not have a strict analogon in quantum theory. Sometimes the set of unit rays in Hilbert space $\mathcal{H}$ is taken as a substitute, but we need not adopt this view here.

We take, as a substitute for μ, a density operator W_0 which is to be given. X^*, Y^*, and Z^* will now denote positive operator valued measures (p.o.v. measures) on $\mathcal{B}(\mathcal{X})$, $\mathcal{B}(\mathcal{Y})$, or $\mathcal{B}(\mathcal{Z})$, respectively, with $X^*(\mathcal{X}) = Y^*(\mathcal{Y}) = Z^*(\mathcal{Z}) = 1$. μ^X shall now denote the probability measure

$$\mu^X := \mathrm{Tr}(X^*(.)\, W_0) \tag{12}$$

on $\mathcal{B}(\mathcal{X})$. μ^Y and μ^Z are defined analogously.

Instead of measurable functions on Ω, or, more precisely, instead of the characteristic functions of their inverse images of Borel sets, we use now the p.o.v. measures. This is somewhat more general since characteristic functions act by multiplication in function space as projection operators but p.o.v. measures need not be projection valued. But the probabilistic interpretation of formula (12) is the same as in the classical case, especially the meaning of the probability space $(\mathcal{X}, \mathcal{B}(\mathcal{X}), \mu^X)$ is maintained.

Nevertheless, our background philosophy by which we have motivated postulates $(\mathrm{I})_c$ and $(\mathrm{II})_c$ for the classical case cannot be taken over in such a simple way. In section 1 we have described observation of $\eta \in \mathcal{B}(\mathcal{Y})$ and selection of the respective subensemble by leaving μ fixed and restricting the range of integration from Ω to $Y_0^*(\eta) \in \Omega$. Y_0 assigns now the classical Y from the preceeding section. In the language of instruments due to E. B. Davies we have performed the operation

$$\mu \mapsto \mu((.) \cap Y_0^*(\eta))\,. \tag{13}$$

But such an operation cannot be formulated in quantum theory. Instead of the latter the projection postulate

$$W_0 \mapsto Y^*(\eta)\, W_0\, Y^*(\eta)$$

has been often considered. But it can be introduced only if the spectrum of the observable to be measured is purely discrete. In general, selection processes have to be described by instruments. Recall that an operation is any positive linear mapping bounded by 1 in the space of real trace class operators. Then an instrument $\mathcal{J}$ on $\mathcal{B}(\mathcal{Y})$ is an operation valued measure on $\mathcal{B}(\mathcal{Y})$ with the property

$$\operatorname{Tr}(\mathcal{J}(\mathcal{Y})W) = \operatorname{Tr}(W)$$

for any positive trace class operator W. If, moreover, for any $\eta \in \mathcal{B}(\mathcal{Y})$ and any W

$$\operatorname{Tr}(\mathcal{J}(\eta)W) = \operatorname{Tr}(Y^*(\eta)W)$$

holds true, then $\mathcal{J}$ is said to measure Y^*. Y^* is uniquely determined by $\mathcal{J}$ but the converse does not hold.

In the quantum case we describe selection processes by some fixed given $\mathcal{J}$ which determines Y^*. Then, up to this point, the background philosophy adopted in section 1 can be literally taken over and we shall not repeat it here. Instead of diagram (1) we have now

$$\begin{array}{ccccc} & \overset{X^*}{\swarrow} & \mathcal{B}(\mathcal{X})\,;\,\mathcal{X} & \xrightarrow{\;f\;} & \mathbb{R} \\ [0,1](\mathcal{H}) & & & & \\ & \underset{Y^*}{\nwarrow} & \mathcal{B}(\mathcal{Y})\,;\,\mathcal{Y} & \xrightarrow{\;E^{\mathcal{J}}(f)\;} & \mathbb{R}\,, \end{array} \tag{14}$$

which has similar probabilistic meaning. $[0, 1](\mathcal{H})$ denotes the order unit interval of bounded hermitean operators in $\mathcal{H}$. $E^{\mathcal{J}}(.)$ is the conditional expectation to be constructed, f being a random variable on $\mathcal{X}$.

The postulate

$$E^{\mathcal{J}}(f) \text{ is a random variable on } \mathcal{B}(\mathcal{Y}) \qquad (\mathrm{I})_q$$

has the same meaning as $(\mathrm{I})_c$. In order to formulate a postulate corresponding to $(\mathrm{II})_c$, remember that composition of measurable functions in classical probability theory has the same meaning as forming

$$f \circ X := \int_{\mathcal{X}} f\, dX^*$$

in functional calculus of selfadjoint operators where X^* is the spectral measure. For X^* being a general p.o.v. measure the latter definition may involve some problems in which we will not enter here. Now the left hand side of $(\mathrm{II})_c$ is the expectation value of $f \circ X$ after selection of the subensemble in which $\eta \in \mathcal{B}(\mathcal{Y})$ has been occured. Transducing this to the quantum case we postulate

$$\operatorname{Tr}((f \circ X) \cdot (\mathcal{J}(\eta) W_0)) = \int_{\eta} E^{\mathcal{J}}(f)\, d\mu^Y . \qquad (\mathrm{II})_q$$

In analogy to (5) we may define

$$\mathbb{E}^{\mathcal{J}}(\eta, f \circ X) := (E^{\mathcal{J}}(f) \cdot \chi_\eta) \circ Y\,,$$

which, in case Y^* is projection valued, is given by

$$\mathbb{E}^{\mathcal{J}}(\eta, f \circ X) = \int_\eta E^{\mathcal{J}}(f)\, dY^*\,. \tag{15}$$

With this definition $(\mathrm{II})_q$ reads as

$$\mathrm{Tr}((f \circ X) \cdot (\mathcal{J}(\eta) W_0)) = \mathrm{Tr}(\mathbb{E}^{\mathcal{J}}(\eta, f \circ X) W_0)\,.$$

$\mathbb{E}^{\mathcal{J}}(\,.\,,\,.)$ has been called an generalized conditional expectation in H. Cycon and K.-E. Hellwig (1977). In two special cases the existence of such an object is proved there. One case is, that W_0 is effective with respect to the observable Y^* which is measured by $\mathcal{J}$. Effectivity of W_0 with respect to Y^* means that $\mathrm{Tr}(Y^*(\eta) W_0) = 0$ implies $Y^*(\eta) = 0$. The construction proceeds just along the lines sketched in section 1. Let f be an integrable function with respect to μ^Y (c. f. formula (12)), then $\mathrm{Tr}((f \circ X) \cdot (\mathcal{J}(.) W_0))$ is a measure on $\mathcal{B}(\mathcal{Y})$ which is subordinated to μ^Y. So $E^{\mathcal{J}}(f)$ can be taken to be a version of the Radon-Nikodym density of $\mathrm{Tr}((f \circ X) \cdot (\mathcal{J}(.) W_0))$ with respect to μ^Y. We have instead of diagram (2)

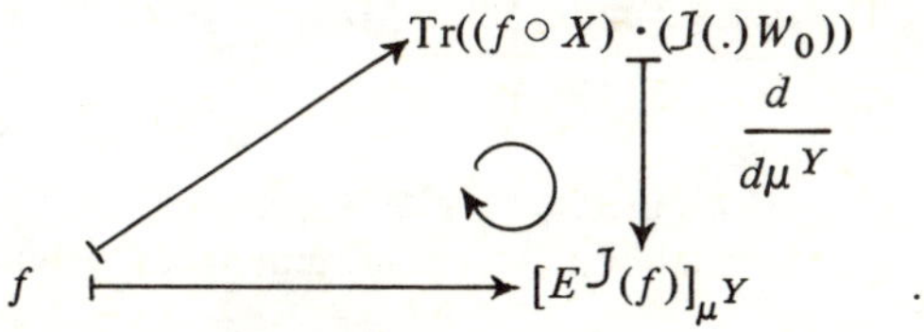

We will now apply the foregoing to the case when Y^* is projection valued and has purely discrete bounded spectrum σ on the real line, i.e. $\mathcal{Y} := \mathbb{R}$. With the notation

$$Y^*_y := Y^*(\{y\}) \tag{16}$$

we have

$$1 = \sum_{y \in \sigma} Y^*_y\,,$$

$$Y := \sum_{y \in \sigma} y\, Y^*_y\,, \tag{17}$$

$$Y^*(\eta) = \sum_{y \in \eta} Y^*_y\,. \tag{18}$$

Moreover, we define the instrument $\mathcal{J}^Y$ measuring Y to be the v. Neumann-Lüders instrument:

$$\mathcal{J}^Y(\eta, W) := \sum_{y\in\eta} Y_y^* \, W \, Y_y^* . \tag{19}$$

Then

$$\mathrm{Tr}((f \circ X) \cdot (\mathcal{J}^Y(\eta) W_0)) = \sum_{y\in\eta} \mathrm{Tr}((f \circ X) \cdot (Y_y^* W_0 Y_y^*))$$

which should (c.f. $(\mathrm{II})_q$) be equal to

$$\int_\eta E^{\mathcal{J}^Y}(f)\, d\mu^Y = \sum_{y\in\eta} (E^{\mathcal{J}^Y}(f))\,(y)\, \mathrm{Tr}(Y_y^* W_0) .$$

Hence, a version of the Radon-Nikodym density is

$$E^{\mathcal{J}^Y}(f) = \sum_{y\in\sigma} \frac{\mathrm{Tr}((f \circ X) \cdot (Y_y^* W_0 Y_y^*))}{\mathrm{Tr}(Y_y^* W_0)} \chi_{\{y\}},$$

where $\chi_{\{y\}}$ is the characteristical function of the one-point set $\{y\}$. The generalized conditional expectation (c. f. formula (12)) is

$$\mathbb{E}^{\mathcal{J}^Y}(\eta, f \circ X) = \sum_{y\in\eta} \frac{\mathrm{Tr}((f \circ X) \cdot (Y_y^* W_0 Y_y^*))}{\mathrm{Tr}(Y_y^* W_0)} Y_y^* .$$

We remark that as in the classical case $\mathbb{E}^{\mathcal{J}^Y}(\,.\,, f \circ X)$ is not the dual object of the instrument $\mathcal{J}^Y$ but the dual of another instrument $\mathcal{K}^Y$ given by

$$\mathcal{K}^Y(\eta) W = \sum_{y\in\eta} \frac{\mathrm{Tr}(Y_y^* W)}{\mathrm{Tr}(Y_y^* W_0)} Y_y^* W_0 Y_y^* ,$$

$\eta \in \mathcal{B}(\mathbb{R})$. The action of this instrument is similar to coarse graining procedure in quantum statistics.

Let now Z be a further observable which is of the same kind as Y such that formulae (16) through (19) hold true replacing Y by Z, y by z, and σ by σ_z. With respect to Z Borel sets on the real line will be denoted by ζ. Selections due to measurements of Z shall be described by the respective v. Neumann-Lüders instrument $\mathcal{J}^Z$. As in the classical case in section 1 we assume Z to be measured prior to Y, and Y prior to X.

When does a Chapman-Kolmogorov equation in the form of formula (10) in section 1 hold true? Inserting the quantal expressions into formula (10) we obtain that

$$\mathbb{E}^{\mathcal{J}^Z}(\{z\}, \mathbb{E}^{\mathcal{J}^Y}(\mathbb{R}, X^*(\xi))) = \sum_{y\in\sigma} \frac{\mathrm{Tr}(X^*(\xi) Y_y^* W_0 Y_y^*) \mathrm{Tr}(Y_y^* Z_z^* W_0 Z_z^*)}{\mathrm{Tr}(Y_y^* W_0) \qquad \mathrm{Tr}(Z_z^* W_0)} Z_z^*$$

should be equal to

$$\mathbb{E}J^Z(\{z\}, X^*(\xi)) = \frac{\mathrm{Tr}(X^*(\xi) Z_z^* W_0 Z_z^*)}{\mathrm{Tr}(Z_z^* W_0)} \; Z_z^* \; .$$

If Z is a function of Y, i.e. in the language of v. Neumann algebras

$$Z \in \{Y\}'' , \tag{20}$$

then let $\varphi : y \longmapsto z$ such that

$$Y_y^* Z_z^* = \begin{cases} Y_y^* , & z = \varphi(y) \\ 0 , & \text{otherwise} \end{cases} ,$$

then

$$\mathbb{E}J^Z(\{z\}, \mathbb{E}J^Y(\mathbb{R}, X^*(\xi))) = \sum_{y \in \varphi^*(\{z\})} \frac{\mathrm{Tr}(X^*(\xi) Y_y^* W_0 Y_y^*)}{\mathrm{Tr}(Z_z^* W_0)} \; Z_z^* \; .$$

If moreover , X commutes with Y, i.e.

$$X \in \{Y\}' , \tag{21}$$

then

$$\sum_{y \in \varphi^*(\{z\})} \mathrm{Tr}(X^*(\xi) Y_y^* W_0 Y_y^*) = \sum_{y \in \varphi^*(\{z\})} \mathrm{Tr}(Y_y^* X^*(\xi) W_0 Y_y^*) =$$

$$\mathrm{Tr}(Z_z^* X^*(\xi) W_0 Z_z^*) = \mathrm{Tr}(X^*(\xi) Z_z^* W_0 Z_z^*) \; .$$

Hence (21) together with (20) imply

$$\mathbb{E}J^Z(\{z\}, X^*(\xi)) = \mathbb{E}J^Z(\{z\}, \mathbb{E}J^Y(\mathbb{R}, X^*(\xi))) \; .$$

Summation over $z \in \zeta$ leads to

$$\mathbb{E}J^Z(\zeta, X^*(\xi)) = \mathbb{E}J^Z(\zeta, \mathbb{E}J^Y(\mathbb{R}, X^*(\xi))) ,$$

which is the analogue of the Chapman-Kolmogorov equation.

At first sight this result is demoralizing. For $(W_0, \mathcal{H}, \{X_s\}_{s \in \mathbb{R}^+})$ be to a Markov process we have the sufficient condition that for any $s_0 \in \mathbb{R}^+$

$$\{X_r\}_{0<r<s_0} \subseteq \{X_{s_0}\}'' , \quad \{X_t\}_{s_0<t<\infty} \subseteq \{X_{s_0}\}' \; .$$

By

$$X_r \in \{X_{s_0}\}'' \subseteq \{X_t\}'$$

for $r < t$ we have that any two elements of the family $\{X_s\}_{s \in \mathbb{R}^+}$ commute. Hence, there is an observable A and a family of functions $\{\psi_s\}_{s \in \mathbb{R}^+}$ such that $X_s = \psi_s(A)$ and the stochastic process reduces to a classical one. Formula (20) means, furthermore, that for $r < t$ X_t is a function of X_r, which means that later measurements have to be more "precise" than earlier ones.

But our assumptions are very special and we have not used the fact that W_0 is fixed in our model. Thus more complicated examples may be possible which are non-trivial quantum Markov processes.

References

H. Cyon, K.-E. Hellwig: Conditional expectations in generalized probability theory, J. Math. Phys. **18**, 1154–1161 (1977).

H. Bauer: Wahrscheinlichkeitstheorie und Grundzüge der Maßtheorie, Walter de Gruyter & Co., Berlin (1974).

E. B. Davies, J. T. Lewis: An operational approach to quantum probability, Commun. math. phys. **17**, 239–260 (1970).

E. B. Davies: Quantum Theory of Open Systmes, Academic Press, London (1976).

C. M. Edwards: The operational approach to algebraic quantum theory I, Commun. math. phys. **16**, 207–230 (1970).

C. M. Edwards: Classes of operations in quantum theory, Commun. math. phys. **20**, 26–56 (1971).

M. Nakamura, T. Turumaru: Expectations in an operator algebra, Tôhoku Math. Journal **6**, 182–188 (1954).

M. Nakamura, M. Takesaki, H. Umegaki: A remark on expectations of operator algebras, Kodai Math. Sem. Rep. **12**, 82–90 (1960).

Shu-Teh Chen Moy: Characterization of conditional expectations as a transformation on function spaces, Pac. J. Math. **4**, 47–64 (1954).

M. D. Srinivas: Foundations of a quantum probability theory, J. Math. Phys. **16**, 1672–1685 (1975).

M. D. Srinivas: Collapse postulate for observables with continuous spectra, preprint (1979).

H. Umegaki: Conditional expectations in an operator algebra, Tôhoku Math. Journal **6**, 177–181 (1954).

H. Umegaki: Conditional expectations in an operator algebra II, Tôhoku Math. Journal **8**, 182–188 (1956).

H. Umegaki: Conditional expectations in an operator algebra III, Kodai Math. Sem. Rep. **11**, 54–64 (1959).

Comparison of the Quantum Logic, Convexity, and Algebraic Approaches to Quantum Mechanics

By

Stanley P. Gudder

1. Introduction

The three main frameworks for studying axiomatic quantum mechanics are the quantum logic, convexity, and algebraic approaches. Although hundreds of papers have been written on each of these three approaches, very little research has been devoted to comparing them. In this article we shall begin such a comparison.

Each of the three approaches has the same goal: to develop a rigorous axiomatic framework for the foundations of quantum mechanics. In each approach, the axiomatic framework is constructed in terms of different primitive undefined elements. Each approach generalizes both the traditional Hilbert space quantum mechanics and classical mechanics so they each reduce to the same structure in these cases. However, in their most general formulation, each approach is different and none includes either of the other two. Fortunately, they contain overlapping structures which allow comparisons.

Each of the three approaches emphasize different aspects of a physical system and each utilizes different mathematical techniques. A comparison allows one to transfer the advantages and techniques of one approach to the others, and aids one in deciding which approach is best for describing a particular situation.

We shall begin with a very general formulation of each approach. To create a standard for comparison, the frameworks we present will frequently be more general than the traditional ones. If one considers stronger axiomatizations then the resulting comparisons will, of course, be different than those given here [20, 21]. In this article we are concerned with the mathematical framework and shall not consider their physical interpretations to a great extent.

2. Quantum Logic Approach

Some of the pioneering investigators in this approach are von Neumann and Birkhoff [2], Piron [18, 19], Jauch [12], Mackey [15], and Varadarajan [24]. Although their physical interpretations may be different, their ultimate mathematical frameworks are quite similar. (This last remark also applies to investigators in the other two approaches).

Primitive Elements. The primitive elements are called *propositions* (*yes-no questions, quantum events*).

Axiomatic Framework. Let $\mathcal{P}$ be a σ-orthocomplete, orthomodular poset whose elements are called *propositions.* A *q-state* is a map $m : \mathcal{P} \to [0, 1]$ such that $m(1) = 1$, and $m(\bigvee_i a_i) = \sum_i m(a_i)$ if $a_i \perp a_j, i \neq j = 1, 2, \ldots$.

A *proposition system* is a σ-orthocomplete, orthomodular poset $\mathcal{P}$ which possesses an order-determining set of q-states.

Derived Elements. A *q-observable* is a σ-homorphism $x : B(\mathbb{R}) \to \mathcal{P}$. If m is a q-state and x a q-observable, the probability measure $m_x : B(\mathbb{R}) \to \mathbb{R}$ defined by $m_x(E) = m[x(E)]$ is the *distribution* of x in the q-state m. The *m-expectation* of x (if it exists) is $m(x) = \int \lambda \, m_x(d\lambda)$. We say that x is *bounded* if $|m(x)| < \infty$ for every q-state m. The *norm* of a bounded q-observable x is $\|x\| = \sup\{|m(x)| : \text{all } m\}$ ($\|x\| < \infty$, if x is bounded). If $u : \mathbb{R} \to \mathbb{R}$ is measurable, we define the q-observable $u(x)$ by $u(x)(E) = x[u^{-1}(E)]$ for all $E \in B(\mathbb{R})$. The *sum* of two bounded q-observables x and y *exists* if there exists a unique q-observable z such that $m(z) = m(x) + m(y)$ for every q-state m. We say that $\mathcal{P}$ satisfies *condition E* if the sum of every pair of bounded q-observables exists. (In general $x + y$ need not exist [7].) Two propositions $a, b \in \mathcal{P}$ are *q-compatible* ($a \leftrightarrow b$) if there exist mutually orthogonal propositions a_1, b_1, c such that $a = a_1 \vee c$, $b = b_1 \vee c$. Two q-observables x, y are *q-compatible* ($x \leftrightarrow y$) if $x(E) \leftrightarrow y(F)$ for all $E, F \in B(\mathbb{R})$.

3. Algebraic Approach

Some of the pioneering investigators in this approach are Jordan, von Neumann, and Wigner [13], Segal [22, 23], Haag and Kastler [11] and Araki [1]. The formulation that we shall give is due to Segal [22] although it is similar to that of Jordan, von Neumann und Wigner. Since the time of this formulation this approach has evolved into the C^*-algebra approach to quantum mechanics [6, 23]. Since some of the axioms of this latter approach have not been physically justified, we shall only consider the earlier formulation.

Primitive Elements. The primitive elements are called (bounded) *observables.*

Axiomatic Structure. An *observable system* is a set of elements $\mathcal{A}$ called *observables* which satisfies:

(A1) $\mathbb{A}$ is a real Banach space with norm $\|\cdot\|$.

(A2) There is an element $I \in \mathbb{A}$ and for every $A \in \mathbb{A}$ and positive integer n there is an element $A^n \in \mathbb{A}$ for which the following holds: if f, g, h are real polynomials and if $f[g(\alpha)] = h(\alpha)$ for all $\alpha \in R$, then $f[g(A)] = h(A)$ where

$$f(A) = \beta_0 I + \sum_{k=1}^{n} \beta_k A^k \quad \text{if} \quad f(\alpha) = \sum_{k=0}^{n} \beta_k \alpha^k$$

(A3) $\|A^2 - B^2\| \leqslant \max(\|A^2\|, \|B^2\|)$, $\|A^2\| = \|A\|^2$.

(A4) $A \mapsto A^2$ is continuous.

Derived Elements. An *a-state* on $\mathbb{A}$ is a linear functional $\omega : \mathbb{A} \to \mathbb{R}$ such that $\omega(I) = 1$ and $\omega(A^2) \geqslant 0$ for all $A \in \mathbb{A}$. Segal [22] has proved that there exists a separating set of a-states. The *symmetrized product* for $A, B \in \mathbb{A}$ is $A \circ B = \frac{1}{2}[(A + B)^2 - A^2 - B^2]$. Although $\circ$ is commutative, it need not be associative, distributive, or real homogeneous. We say that $A, B \in \mathbb{A}$ are *a-compatible* if $(A \circ C) \circ B = A \circ (C \circ B)$ for every $C \in \mathbb{A}$. In order to get stronger results, the assumption is frequently made that $\mathbb{A}$ is isomorphic to an observable system of self-adjoint elements in a C^*-algebra [6, 23]. A necessary condition for such an embedding is that $\circ$ be distributive (we then call $\mathbb{A}$ a *distributive* observable system). It is not known whether this condition is sufficient.

Lemma [9]. The following statements are equivalent:
(a) A and $-I$ are compatible for every $A \in \mathbb{A}$.
(b) $\circ$ is distributive.
(c) $\circ$ is real homogeneous.

4. Convexity Approach

Some of the pioneering investigators in this approach are Ludwig [14], Mielnik [16. 17], Davies [3, 4], Lewis [4], and Edwards [5]. Although the formulation given here begins more generally, it is equivalent to some of the above frameworks.

Primitive Elements. The primitive elements are called *states.*

Axiomatic Framework. A *convex prestructure* is a nonempty set S together with a map from $[0, 1] \times S \times S$ into S denoted by $(\lambda, s, t) \mapsto \langle\lambda, s, t\rangle$. An *affine functional* is a map $f : S \to \mathbb{R}$ such that

$$f(\langle\lambda, s, t\rangle) = \lambda f(s) + (1 - \lambda) f(t).$$

for every $s, t \in S$, $\lambda \in [0, 1]$. The elements of S are called *normalized states* and the set of affine functionals on S is denoted by S^*. A *state system* is a convex prestructure S for which S^* is separating.

Derived Elements. The *quantum counter* is the functional $\tau \in S^*$ defined by $\tau(s) = 1$ for every $s \in S$. Define the metric σ on S by $\sigma(s, t) = \inf\{0 \leqslant \lambda \leqslant 1 : \langle \lambda, s_1, s\rangle = \langle \lambda, t_1, t\rangle, s_1, t_1 \in S\}$. The metric $\rho(s, t) = \sigma(s, t) / [1 - \sigma(s, t)]$ is the *intrinsic* metric. An element $f \in S^*$ is an *effect* if $0 \leqslant f(s) \leqslant 1$ for all $s \in S$. We interpret $f(s)$ as the probability that the effect f is observed if the system is in the normalized state s. Denote the set of effects by $\mathcal{E}(S)$. A *c-observable* is a map $x : B(\mathbb{R}) \to \mathcal{E}(S)$ such that $x(\mathbb{R}) = \tau$, $x(\cup E_i) = \sum_i x(E_i)$, $E_i \cap E_j = \phi, i \neq j = 1, 2, \ldots$, where the convergence is pointwise.

Theorem [8]. *S is isomorphic to a generating convex set $\hat{S}$ in a real normed linear space $(\mathcal{V}(S), \|\cdot\|)$. Identify S and $\hat{S}$ and let K denote the cone generated by $\hat{S}$ in $\mathcal{V}(S)$. Then $\|s - t\| = \rho(s, t)$ for all $s, t \in S$, and τ has a unique continuous linear extension $\hat{\tau}$ to $\mathcal{V}(S)$ such that $\hat{\tau}(p) = \|p\|$ for all $p \in K$. Any effect f on S has a unique extension to a positive linear functional $\mathcal{V}(S)$ with $\|f\| \leqslant 1$.*

We call K a *state space* and the elements of K are called *states.* It follows from the theorem that $(\mathcal{V}(S), \|\cdot\|)$ is a base normed space [4]. An *operation* is a positive linear operator $T : \mathcal{V}(S) \to \mathcal{V}(S)$ such that $\hat{\tau}[Tp] \leqslant \hat{\tau}(p)$ for all $p \in K$. Denote the set of operations by $\mathcal{O}(S)$. An *instrument* is a map $\mathcal{D} : B(\mathbb{R}) \to \mathcal{O}(S)$ such that $\tau[\mathcal{D}(\mathbb{R})p] = \tau(p)$ for all $p \in K$ and $\mathcal{D}(\bigcup_i E_i) = \sum_i \mathcal{D}(E_i)$, $E_i \cap E_j = \phi$, $i \neq j = 1, 2, \ldots$, where the sum converges in the strong operator topology.

5. Comparisons

We now give a method for comparing the three approaches. It turns out that there are six possible comparisons, each comparison involving an embedding of one of the approaches into another. Most of these comparisons form major projects with much work yet to be done. We shall briefly mention each of the six comparisons and give some of the unsolved problems that need investigation.

Quantum Logic → Algebraic. Let $\mathcal{P}$ be a proposition system and let $B(\mathcal{P})$ be the set of all bounded q-observables on $\mathcal{P}$. The first question we might ask is whether $B(\mathcal{P})$ is isomorphic to a subset of an observable system. The author does not know the answer to this question in general. However, if $\mathcal{P}$ satisfies condition E, then using the norm $\|\cdot\|$ defined above for $x \in B(\mathcal{P})$, it can be shown that $(B(\mathcal{P}), \|\cdot\|)$ is a real normed linear space and its completion $\mathcal{A}(\mathcal{P})$ is a distributive observable system. Moreover, the q-states on $\mathcal{P}$ have unique extensions to a-states on $\mathcal{A}(\mathcal{P})$. Also, $x, y \in B(\mathcal{P})$ are q-compatible if and only if they are a-compatible. For $a \in \mathcal{P}$, define $x_a \in B(\mathcal{P})$ by $x_a(\{1\}) = a$, $x_a(\{0\}) = a'$. Then x_a is idempotent; that is, $x_a^2 = x_a$. Let $\hat{\mathcal{P}}$ denote the set of idempotent elements of $B(\mathcal{P})$. For $A, B \in \hat{\mathcal{P}}$, define $A \leqslant B$ if $\omega(A) \leqslant \omega(B)$ for every a-state ω and define $A' = I - A$. Then $\hat{\mathcal{P}}$

is a σ-orthocomplete orthomodular poset and $a \mapsto x_a$ is an isomorphism of P onto $\hat{P} \subseteq A(P)$. In this sense, we can say that a proposition system satisfying condition E can be embedded in an observable system.

Algebraic → Quantum Logic. Can an observable system A be embedded in $B(P)$ where P is a proposition system satisfying condition E? Motivated by the previous paragraph, we consider the set D of idempotents in A. Unfortunately, D can be very small. For example, in the C^*-algebra of continuous functions on the unit interval we have $D = \{0, I\}$. In any case we have

Lemma [10]. D is an orthocomplemented poset.

One can give examples of observable systems in which D is not orthomodular [10]. Such examples can be used to show that one can not embed A into a $B(P)$ in general. However, such examples are nondistributive. It is not known if a distributive A can be embedded in a $B(P)$. We do have the following weaker result.

Theorem [10]. *A can be embedded in $B(P)$ for a σ-orthocomplete orthocomplemented poset P.*

Although the above embedding preserves the operations in A it need not preserve states and compatibility.

Quantum Logic → Convexity. In this case we have some satisfactory answers. Let P be a proposition system and let S be the set of all q-states on P. Then S is closed under convex combinations $(\lambda, s, t) \mapsto \lambda s + (1-\lambda)t$ so S is a convex prestructure. If $a \in P$, then $f_a : S \to \mathbb{R}$ defined by $f_a(s) = s(a)$ is affine. In fact, f_a is an effect since $0 \leqslant f_a(s) \leqslant 1$. Moreover, $f_1 = \tau$. If $s_1 \neq s_2 \in S$, then there exists an $a \in P$ such that $s_1(a) \neq s_2(a)$ so $f_a(s_1) \neq f_a(s_2)$. Hence, S is a state system. If $Ep(S)$ denotes the set of convex combinations of f_a's, then $Ep(S)$ is a convex subset of S^*. Let $\bar{P} = \{f_a : a \in P\} \subseteq Ep(S)$. Then $a \mapsto f_a$ is a bijection from P to $\bar{P}$.

Lemma. If S is strongly order determining, then $\bar{P}$ is the set of extremal elements of $Ep(S)$.

In $E(S)$, define $f' = \tau - f$ and $f \leqslant g$ if $f(s) \leqslant g(s)$ for every $s \in S$. It can then be shown that $\bar{P}$ is a σ-orthocomplete orthomodular poset and P and $\bar{P}$ are isomorphic. We then obtain the following result.

Theorem. *A proposition system can be embedded into a state system.*

Furthermore, this embedding preserves observables. In fact, if x is a q-observable and we define $\bar{x} : B(\mathbb{R}) \to E(S)$ by $\bar{x}(E) = f_{\bar{x}(E)}$, then $\bar{x}$ is a c-observable.

Convexity → Quantum Logic. If S is a state system, is S isomorphic to a set of states on a σ-orthocomplete orthomodular poset? This does not seem possible in a physically reasonable way. For example, one might consider the set of extremal effects $E_1(S)$ on S. For $f \in E_1(S)$, define $f' = \tau - f$.

Then $f' \in E_1(S)$. For $f, g \in E_1(S)$, define $f \leqslant g$ if $f(s) \leqslant g(s)$ for all $s \in S$. Then $\leqslant$ is a partial order relation on $E_1(S)$ with first and last elements $0, \tau$. Clearly, $f'' = f$ and $f \leqslant g$ implies that $g' \leqslant f'$. However, the other axioms do not seem to hold.

Algebraic → Convexity. Very little seems to be known in this case. Let $\mathcal{A}$ be an observable system and let S be the set of a-states on $\mathcal{A}$. If $A \in \mathcal{A}$ is positive ($A = B^2, B \in \mathcal{A}$) and $\|A\| \leqslant 1$, define $f_A(s) = s(A)$ for all $s \in S$. Then $f_A \in E(S)$ and these effects separate states so S is a state system. However, it is not clear how one should embed the elements of $\mathcal{A}$ into the set of c-observables on S.

Convexity → Algebraic. It is not clear how one can equip the observables of a state system with the algebraic structure of an observable system. Even in the traditional Hilbert space approach, observables are P 0 V measures on a Hilbert space and sums of observables seem to have no physical meaning. In the traditional approach, S is the set of positive trace class operators and the dual of S is the set of bounded self-adjoint operators which correspond to the extremal observables. In the general case, however, the dual of $V(S)$ need not be an algebra of operators.

References

[1] H. Araki, On the algebra of all local observables, Progr. Theoret. Phys. **32** (1964), 844–854.

[2] G. Birkhoff and J. von Neumann, The logic of quantum mechanics, Ann. Math. **37** (1936), 823–843.

[3] E. Davies, Quantum Theory of Open Systems, Academic Press, New York (1976).

[4] E. Davies and J. Lewis, An operational approach to quantum probability, Commun. Math. Phys. 17 (1970), 239–260.

[5] C. Edwards, The operational approach to algebraic quantum theory I, Commun. Math. Phys. 16 (1970), 207–230.

[6] G. Emch, Algebraic methods in Statistical Mechanics and Quantum Field Theory. Wiley-Interscience, New York (1972).

[7] S. Gudder, Uniqueness and existence properties of bounded observables, Pac. J. Math. 19 (1966), 31–93. [Correction, Pac. J. Math. 19 (1966), 588–589.]

[8] S. Gudder, Stochastic Methods in Quantum Mechanics, North Holland, New York (1979).

[9] S. Gudder, Algebraic conditions for a function on an abelian group, Letts. Math. Phys. 3 (1979), 127–133.

[10] S. Gudder and S. Boyce, A comparison of the Mackey and Segal models for quantum mechanics, Intern. J. Theoret. Phys. 3 (1970), 7–21.

[11] R. Haag and D. Kastler, An algebraic approach to quantum field theory, J. Math. Phys. 5 (1964), 848–861.

[12] J. Jauch, Foundations of Quantum Mechanics, Addison-Wesley, Reading, Mass. (1968).

[13] P. Jordan, J. von Neumann, and E. Wigner, On an algebraic generalization of the quantum mechanical formalism, Ann. Math. 35 (1934), 29–64.

[14] G. Ludwig, Attempt of an axiomatic foundation of quantum mechanics and more general theories III, Commun. Math. Phys. 9 (1968), 1–12.
[15] G. Mackey, Mathematical Foundations of Quantum Mechanics, Benjamin, New York (1963).
[16] B. Mielnik, Geometry of quantum states, Commun. Math. Phys. 9 (1968), 55–80.
[17] B. Mielnik, Theory of filters, Commun. Math. Phys. **15** (1969), 1–46.
[18] C. Piron, Axiomatique quantique, Helv. Phys. Acta **37** (1964), 439–468.
[19] C. Piron, Foundations of Quantum Physics, Benjamin, Reading, Mass. (1976).
[20] R. Plymen, C*-algebras and Mackey's axioms, Commun. Math. Phys. 8 (1968), 132–146.
[21] R. Plymen, A modification of Piron's axioms, Helv. Phys. Acta **41** (1968), 69–74.
[22] I. Segal, Postulates for general quantum mechanics, Ann. Math. **48** (1947), 930–948.
[23] I. Segal, Mathematical Problems of Relativistic Physics, Amer. Math. Soc., Providence, R. I. (1963).
[24] V. Varadarajan, Geometry of Quantum Theory I and II, Van Nostrand, Princeton, N. J. (1968, 1970).

Connections between Different Approaches to the Foundations of Quantum Mechanics

by

G. Ludwig and H. Neumann

In this contribution we investigate the connections between different approaches to the foundation of quantum mechanics. We shall be concerned with the following concepts:

FQP The foundation of quantum theory according to C. Piron as described in [1] and [2].

OS The operational statistics approach developed by C. H. Randall and D. J. Foulis [3], [4] (this volume).

LQP The propositional language approach according to P. Mittelstaedt and E.-W. Stachow as discussed in [5], [6] (this volume).

ABQM The axiomatic basis of quantum mechanics developed by one of the authors. A survey of this approach is given in [7] (this volume).

In the following three sections we shall study mainly the relationship of ABQM to the other concepts and thus the connections among FQP, OS and LQP are exhibited indirectly only. Moreover, the discussion concerning LQP is of a preliminary character and mainly points out some outstanding problems.

I. Connections between FQP and ABQM

Since both of the theories considered have a *realistic* background it is not so difficult to find relations between them. In ABQM the starting point of the theory is the reality of the preparation and registration procedures, a reality given by "pretheories". In this sense the basic reality is a phenomenological one. With the construction of the theory the more extensive reality domain of the microsystems is deduced from the basic realities according to the general methods developed in [8].

Piron's approach [FQP] has also a realistic background. This can be seen not only from [1] but also in a more impressive form from a lecture given on a "Symposion on Logic, Measurement and Probability" in Reisensburg [9]. We quote:

"By a physical system we understand a part of the reality conceived as existing in space-time and exterior to the physicist. The precise definition of a physical system always depends on the point of view and the degree of idealization considered. As we have said we adopt the realistic point of view. The system is and it is what it is. It possesses different properties and whether these are known or not by the physicist does not change anything to the reality of the object itself. We have to describe these properties and not to justify or to explain the knowledge of the physicist about them".

This concept of "system" is applied to those systems which are called microsystems in ABQM. Therefore, in the following we shall use the words "system" in FQP and "microsystem" in ABQM as describing the same "parts of reality".

We are now looking to introduce something in ABQM corresponding to what Piron calls "properties". We will be guided by the following explanations of Piron in [9].

"When we believe that the system has actually a certain property we may decide to test our believe by performing a suitable experiment. If we do this and if we obtain what we are expecting we are reinforced in our believe but if we obtain something else we must admit our error and change our idea. In other words we cannot and we never prove a theory we only have the possibility to falsify it. That is why a physical property must always be associated to some experiment and a well-defined result.

If one given system has been prepared in such a way that we can affirm that in the event of the experiment the expected result would be certain, we will say that the corresponding property is an actual property of the system in opposition to the other properties which are only potential.

The first problem in physics is to exhibit a mathematical structure able to describe the collection of all actual and potential properties of a system. To do this we proceed on the following way."

And we will follow this way; Piron continues:

"We first define the concept of question. A question is an experiment leading to a well defined alternative of which the terms are "yes" or "no", the "yes" corresponding to the expected result and the "no" to all the others."

Therefore the first step on the way to introduce the concept of a property is the introduction of the concept of question in ABQM. For this concept a more elaborate definition is given by Piron in [2]:

"*Définition:* On appelle *question* toute expérience conduisante à une alternative dont les termes sont "oui" ou "non". Plus schématiquement, pour définir une question, il faut se donner:

1) Le ou des appareils.

2) Leurs modes d'emploi.

3) Une règle univoque permettant d'interpréter le resultat quel qu'il soit: le "oui" correspondant au résultat cherché, le "non" à tous les autres."

Piron denotes by Q the set of questions. We will write here Q_p. Correspondingly we will use the same letters as in [1], [2] adding, however, an index P to distinguish these letters from those used in ABQM.

The definition above shows that the physical interpretation of the set Q_P of questions and of the set $\mathcal{F}$ of effect procedures, given in ABQM, is exactly the same. So one would be inclined to identify Q_P and $\mathcal{F}$. However, we will see later that Q_P includes also "idealized" effect procedures and thus must be larger than $\mathcal{F}$. Therefore we introduce a set $\hat{\mathcal{F}} \supset \mathcal{F}$ in ABQM such that the mapping $\psi : \mathcal{F} \to \mathcal{L}$ ([7], § 3) can be extended to a surjective mapping $\psi : \hat{\mathcal{F}} \to L$ (L as in [7] § 4). We will now identify $\hat{\mathcal{F}}$ and Q_P in the following sense: The elements of $\hat{\mathcal{F}}$ and Q_P have the same physical interpretation *and* the set $\hat{\mathcal{F}}$ in ABQM is a term of the same structure as Q in FQP (see also in general the comparison of physical theories and the concept of equivalent theories as described in [8] § 8).

We assume that also the elements of $\hat{\mathcal{F}}$ have the form (b_0, b) with subsets $b_0, b \subset M$ (however, $b_0 \in \hat{R}_0 \supset R_0$ and $b \in \hat{R} \supset R$). Piron introduces the "inverse question" $\tilde{\alpha}$ for all $\alpha \in Q_P$. Evidently, the relation $\alpha \mapsto \alpha$ in Q_P corresponds to the relation $(b_0, b) \mapsto (b_0, b_0 \setminus b)$ in $\hat{\mathcal{F}}$.

Another relation introduced in FQP is the product $\pi\alpha_i$ of questions α_i. We refer to the definition given in [2]:

"Si α_i $i \in J$ est une famille de questions, on peut définir $\underset{j}{\pi}\, \alpha_i$, la question produit obtenue de la manière suivante:

1) On se donne l'ensemble des appareils et des modes d'emploi correspondant aux α_i.

2) Pour exécuter l'expérience correspondante à $\underset{i}{\pi}\alpha_i$, on doit, par définition, procéder ainsi: on choisit librement (ou au hasard) l'une des questions α_i et on exécute l'expérience correspondante.

3) On attribue à $\underset{i}{\pi}\, \alpha_i$ la réponse "oui" ou "non" ainsi obtenue."

(Since $\underset{i}{\pi}\, \alpha_i \in Q_P$ for all $\{\alpha_i\}_{i \in J}$ the set Q_P can't be denumerable in contrast to $\mathcal{F}$. This is one of the reasons to enlarge $\mathcal{F}$ to $\hat{\mathcal{F}}$.)

It is not difficult to see that the product $\underset{i}{\pi}\, \alpha_i$ in Q_P corresponds to the direct mixture of the corresponding (b_{oi}, b_i) in $\hat{\mathcal{F}}$. The definition of the direct mixture of recording methods is given in [10] III, § 2.

Thereby it is necessary to mention that the definition of the direct mixtures comprises the determination of the statistical weights of the selection of the different α_i (i.e. (b_{oi}, b_i)) by chance ("au hasard"). This is not necessary in FQP because the "certainty" of a question, alone, is of interest as we shall see later.

In order to introduce the notion of "properties" we quote [9]:

"To each question corresponds a property of the system and we will say that the property is actual or that the question is true if the physical system is actually such that in the case where we would perform the question the answer "yes" would be certain."

In [2] Piron gives the precise definition:

"*Définition:* Si le système physique est tel que si on effectuait la question $\alpha(\alpha \in Q)$, la réponse "oui" serait certaine, on dit que la question α est *vraie* actuellement pour ce système et ceci même si effectivement on n'a pas du tout l'intention d'exécuter actuellement l'expérience correspondante."

Before translating the notion of properties to ABQM we have to translate the concept of truth of questions. To this end we introduce a formal relation $w(x,f)$ for $x \in M$ and $f \in \hat{F}$ with the physical interpretation as described in definition above, i.e. $w(x, f)$ is the mathematical picture of: f is true for the system x. What is the connection between this formal relation and other ones already defined in ABQM? The above given quotation:

"If one given system has been prepared in such a way that we can affirm that in the event of the experiment the expected result would be certain, we will say that the corresponding property is an actual property of the system. . ."

gives us a hint for the wanted connection: $w(x, f)$ means that x has been prepared in such a way that $f = (b_o, b)$ is certain for x, i.e. there is an $a \in Q$ with $x \in a$ and $\lambda(a \cap b_o, a \cap b) = 1$. Again we are careful and admit "idealized" preparing procedures in identifying these relations: We introduce a set $\hat{Q} \supset Q$ in ABQM such that the mapping $\varphi : Q \to K$ can be extended to a surjective mapping $\varphi : \hat{Q} \to K$ (K as in [7] § 4).

For $f = (b_o, b) \in \hat{F}$ we define

$$\hat{Q}_1(f) = \{a : a \in \hat{Q},\ \lambda(a \cap b_o, a \cap b) = 1\} \text{ and } M_1(f) = \bigcup_{a \in \hat{Q}_1(f)} a.$$

and identify $w(x, f)$ with $x \in M_1(f)$.

The preorder relation $\alpha_1 < \alpha_2$ defined on Q_P in FQP corresponds to $M_1(f_1) \subset M_1(f_2)$ ($f_1, f_2 \in \hat{F}$ corresponding to $\alpha_1, \alpha_2 \in Q_P$) and the resulting equivalence relation $\alpha_1 \sim \alpha_2$ on Q_P is to be translated to $M_1(f_1) = M_1(f_2)$. Thus the ordered set L_P of propositions introduced in FQP is order isomorphic to the set $\{M_1(f) : f \in \hat{F}\}$ endowed with the usual set ordering. The elements of the set $\{M_1(f) : f \in \hat{F}\}$ in ABQM are representatives of the properties in the sense of Piron and the set of all "actual properties" of a system x is then (translated to ABQM) the set $\{M_1(f) : f \in \hat{F} \text{ and } x \in M_1(f)\}$.

With $\lambda(a \cap b_o, a \cap b) = \mu(\varphi(a), \psi(b_o, b))$
and $K_1(\psi(f)) = \{W : W \in K, \mu(W, \psi(f)) = 1\}$ ([7], § 5),

we may write $M_1(f) = \bigcup_{\substack{a\in Q \\ \varphi(a)\in K_1(\psi(f))}} a.$

Since $\varphi : \hat{Q} \to K$ and $\psi : \hat{F} \to L$ are assumed to be surjective the mapping $K_1(\psi(f)) \longmapsto M_1(f)$ is an order isomorphism of $\{M_1(f) : f \in \hat{F}\}$ and the set $\{K_1(g) : g \in L\}$ of all exposed faces of K ([7], § 5).

It turns out that a state of a system x being defined in FQP as the set of all the actual properties of x corresponds to a maximal filter of the lattice $\{K_1(g) : g \in L\}$. A complete mathematical correspondence is gained if in ABQM atomicity of the lattice $\{K_1(g) : g \in L\}$ is assumed as it is generally assumed for L_P in FQP. The states of FQP then correspond to the atoms of $\{K_1(g) : g \in L\}$ (and the maximal filters they generate). If enough axioms are introduced in ABQM the atoms of $\{K_1(g) : g \in L\}$ can be proved to consist of the extreme points of K.

Having thus accomplished a translation between FQP and ABQM we can f.e. formulate the axioms and theorems of FQP as relations in ABQM and check whether they are theorems in ABQM. It is not difficult to see that this is the case after having introduced all the axioms of ABQM. It could be a future task to investigate which axioms of ABQM are necessary and sufficient to prove axioms and theorems of FQP.

2. Connections between OS and ABQM

In contrast to the concepts of FQP and ABQM the operational statistics approach is no attempt to give an axiomatic foundation of quantum theory. Similar to mathematical statistics it is rather a development and analysis of a mathematical structure which can be applied to many physical theories.

This fact becomes particularly clear in a discussion of the interpretation of the basic set X of "outcomes" in OS. The description of the interpretation in [3] section 2 allows for a broad field of possibilities. Though the expression "outcome" suggests a connection with concrete signals of experimental apparata these connections can be extremely complicated. (This comes particularly apparent in the later discussion of the Hilbert space manuals.) The OS approach is no complete physical theory in the sence that no general rules are provided how to assign the symbols used for outcomes to concrete signals of apparata. Hence, OS is rather a framework or language for a physical theory. This aspect of OS is supported by the general remarks in section 2 of [3] which need not be quoted here.

In the sequel we shall illustrate OS by applying the notions developed in this framework to structures defined in ABQM. With ABQM we shall define several manuals of operations and thus introduce different interpretations for the outcomes and the related notions.

Firstly, we shall discuss the connection between the structure of statistical selection procedures and operational statistics. As operational statistics, the structure of statistical selection procedures introduced in ABQM also generalizes Kolmogorov's foundation of statistics. The generalization in ABQM consists mainly in using only relative probabilities. This characterizes also a difference between ABQM and OS. Hence, in order to gain a simple connection between selection procedures and operations we introduce an additional assumption in the concept of selection procedures. This assumption will not imply contradictions in connection with axioms leading to the quantum structure of ABQM. However, within the framework of ABQM they are unnecessarily restrictive.

Let S be a structure of species 'selection procedures' ([7], § 1). For the following considerations we assume

(D) There is a subset $S_m \subset S$ with $a \cap a' = \phi$ for all $a, a' \in S_m$ and for all $a \in S$ there is $a' \in S_m$ with $a \subset a'$.

As a result of assumption (D) S is a collection of mutually disjoint Boolean algebras of sets $S_a = \{a' \in S : a' \subset a\}$, $a \in S_m$. Assumption (D) appears rather natural if one keeps in mind that for $a, a' \in S$ the relation $a \subset a'$ can be interpreted as: An apparatus realizing the selection procedures a' allows to discriminate the systems (e.g. via signals) according to the procedure a.

This interpretation also suggests the definition of operations according to OS: Operations are finite partitions of maximal elements of S; i.e. with S as the set of outcomes we define the manual of operations corresponding to S by

$$\mathcal{A}_S = \{\{a_i\}_{i=1,\ldots,n} : \phi \neq a_i \in S, a_i \cap a_k = \phi \text{ for } i \neq k, \bigcup_{i=1}^{n} a_i \in S_m\}.$$

The corresponding operational logic is given by $L = \dot{\bigcup}_{a \in S_m} S_a$ where $\dot{\cup}$ denotes the disjoint union of the Boolean algebras S_a under identification of the 0- and 1-elements. (L is no Boolean algebra but only an operational logic according def. 4.2 of [3].)

Now, let (S, λ) be a structure of species "statistical selection procedures" with S satisfying (D). We note that for $a \in S$ fixed $\lambda(a, a')$ is an effective measure on the Boolean algebra S_a. Since for each $a' \in S, a' \neq \phi$, there is exactly one $a \in S_m$ with $a' \subset a$, a weight (def. 5.1 of [3]) $\omega(a') = \lambda(a, a')$ is defined on $\mathcal{A}_S$. Conversely, every effective weight ($\omega(a) = 0 \Rightarrow a = \phi$) on $\mathcal{A}_S$ defines a statistical function $\lambda(a, a') = \omega(a')/\omega(a)$ for $a', a \in S$, $a' \subset a \neq \phi$.

The manual $\mathcal{A}_S$ corresponding to a system of statistical selection procedures describes an essentially classical situation (comp. section 3 of [4]).

In ABQM the structure of selection procedures and statistical selection procedures are applied in various situations. The set Q describing the preparing procedures and R_0 describing the registration methods are statistical selection procedures. The registration methods are statistical selection procedures just insofar as they are not influenced by the preparing procedures. The influence of the preparing procedures is described by enlarging R_0 to R, where R is of the species selection procedures, and by assuming that there is a statistical function λ_S on the set of selection procedures S generated by $\theta = \{a \cap b : a \in Q, b \in R\}$. If Q, R_0 are assumed to satisfy (D), S fulfills (D) with $S_m = \{a \cap b_0 : a \in Q_m, b_0 \in R_{om}, a \cap b_o \neq \phi\}$.

To establish a connection between A_S and the product manuals defined in [3] and [4] consider $A_Q \times A_R$ with the outcome space $Q \times R$ (def. 8.1 of [3]). The mapping $\vartheta(a, b) = a \cap b$ defines a mapping $\vartheta : Q \times R \to S$ which maps products of partitions of maximal elements onto partitions of maximal elements of S or onto empty sets. (Compare the slightly different condition (i) for interpretation morphisms in def. 6.1 of [3]. Condition (ii) is fulfilled.) For a weight ω on A_S, in general, $\omega \circ \vartheta$ is no weight on $A_Q \times A_R$ since $\omega \circ \vartheta$ doesn't sum to 1 for partitions of $a \in Q_m$, $b_o \in R_{om}$ if $a \cap b_o = \phi$.

If the problem of combinability of preparing procedures and recording methods is neglected we may assume $a \cap b_0 \neq \phi$ for all $a \in Q_m$, $b_0 \in R_{om}$. In this case ϑ is an interpretation morphism and $\omega \circ \vartheta$ is a weight on $A_Q \times A_R$ for a weight ω on A_S. It is easily seen that the statistical function λ_S on S satisfies axiom APS 7 of [7] if and only if the corresponding state $\omega \circ \vartheta$ on $A_Q \times A_R$ is a product state if restricted to $A_Q \times A_{R_o}$; i.e. λ_S exhibits no correlations between events of Q and R_0. The construction of the manual $\overrightarrow{A_Q A_R}$ (def. 8.2 of [3]) assures that the manual of operations A_R $(= R_m)$ has no observable influence on A_Q $(= Q_m)$ with respect to states on $\overrightarrow{A_Q A_R}$ (sect. 2 of [4]). According to our statement above $\overrightarrow{A_Q A_R}$ does not completely describe the situation (Q, R_0, R) since in addition there are no correlations between events of Q and R_0.

With these remarks we conclude the discussion of the connection between operational statistics and statistical selection procedures.

Because of the properties of λ_S just mentioned, in ABQM it proves sufficient to consider only the function μ defined on $Q \times F$ by $\mu(a, f) = \lambda_S(a \cap b_0, a \cap b)$ where $f = (b_0, b) \in F$ ([7], § 3).

As in section 1 of this contribution we assume that the sets Q, R_0, R can be completed to $\hat{Q}$, $\hat{R}_0$, $\hat{R}$ by adding idealized procedures such that the mappings $\varphi : \hat{Q} \to K$ and $\psi : \hat{F} \to L$ are surjective. Correspondingly we assume that $\sum_{i=1}^{\infty} \lambda_S(a \cap b_0, a \cap b_i) = 1$ also for infinite partitions of

$$b_0 = \bigcup_{i=1}^{\infty} b_i, \; b_i \in R.$$

By means of the sets $\hat{F}$ and L it is also possible to define certain manuals of operations and study their relations. Firstly, also without assumption (D), we define a manual A_1 with outcome set $\hat{F} = \{(b_0, b) : b_0 \in \hat{R}_0, b \in \hat{R}, b \subset b_0\}$. Let $A_1 = \{b_0 \times \{b_i\}_{i=1,\dots} : b_0 \in \hat{R}_0, \phi \neq b_i \in \hat{R}, b_i \cap b_k = \phi$ for $i \neq k$, $b_0 \bigcup_i b_i\}$. As the manuals discussed above A_1

describes an essentially classical situation. Each preparing procedure $a \in \hat{Q}$, $a \neq \phi$, defines a weight on $\mathcal{A}_1$.

By $\mathcal{A}_2 = \{\{g_i\}_{i=1,\dots} : g_i \in L, \sum_i g_i = 1\}$ another manual of operations with outcome set L is defined. $\mathcal{A}_2$ is not orthocoherent. Note that $\mathcal{A}_2 \subset \mathcal{P}(L)$ implies $g_i \neq g_k$ if $i \neq k$ for an operation $\{g_i\}_{i=1,\dots}$ and therefore $\psi : \hat{\mathcal{F}} \to L$ is no interpretation morphism.

We get an orthocoherent manual by the restriction to the subset $G \subset L$ of outcomes, $G = \partial_e L$: $\mathcal{A}_3 = \{\{e_i\}_{i=1,\dots} : e_i \in G, \sum_i e_i = 1\}$. The operational logic of $\mathcal{A}_3$ is the quantum logic given by G and, considering only the subset $\psi^{-1}(G) \subset \hat{\mathcal{F}}$ of outcomes for $\mathcal{A}_1$, ψ is an interpretation morphism.

The further restriction to the set AG of atoms of G gives the so-called frame manual $\mathcal{A}_4 = \{\{e_i\}_{i=1,\dots} : e_i \in \text{AG}, \sum_i e_i = 1\}$ in case of an irreducible quantum system, where G is the set of projection operators of a Hilbert space. $\mathcal{A}_4$ induces the same logic as $\mathcal{A}_3$. These examples illustrate the variety of physical interpretations which can be given to the set of outcomes of a manual of operations. As we pointed out already in the beginning the connection between concrete signals of apparata and outcomes of a manual may be rather complicate. We quote [3]:

"The decision as to just which outcomes are to be regarded as being physically equivalent could be based on a variety of considerations ranging from an attractive whim to an elaborate scientific theory – often such a decision is inspired by a particular world picture or model."

3. Connections between LQP and ABQM

The propositional language approach considered as a development of a formal language and logic has no direct connection to the theories FQP, OS or ABQM discussed above. In order to get a physically meaningful language a formal language must be applied. This problem of application of the language provides the possibility for relating LQP to the other theories. However, to discuss the resulting connection thoroughly the problem of application of the formal language seems to be not yet elaborate enough. Therefore we shall only try to give some comments.

The problem of application is described for instance in [11] as follows:

"*Elementary propositions* belong to the set of atomic sentences of the syntax of the language. They are predictions of particular properties of a quantum mechanical system. We assume that elementary propositions can be proved or disproved by means of operational tests. Let us consider the following example. An electron (S) is given which moves in the +z-direction of a coordinate system. An elementary proposition is: "The electron has 'spin-up' in the direction a in the $x-y$ plane at the time t", in symbols $a(S, t)$, where t is the future point of time with respect to which the property is predicted. An operational test of this proposition could be a Stern-Gerlach experiment, which is oriented in the direction a together with a photographic plate. The mark, which the electron leaves on the plate, indicates the deviation of the electron in the magnetic field and, thus, proves or disproves the proposition under consideration. Of course, this description of an operational test is a rather simplified and imprecise description. A careful and detailed analysis of a physical operation, provided with a mathematical model, is given by Randall and Foulis. However we accept the view that a precise account of what a physical operation means can be given only in the context of an established physical theory, or in the context of an already established language of physics. Hence we are led to the following commitment. The operational proof procedures for elementary propositions, which belong to the underlying semantics of the language to be established, must be described precisely and completely by the language in the final stage when the language is established. This demand with respect to an operational approach to a physical theory is called *semantical consistency* (v. Weizsäcker) or *self-consistency* (Mittelstaedt.) We cannot here investigate the difficult problem of the consistency of the concept of elementary propositions with respect to an underlying semantics. Since we are, in the first place, interested in a formal logical system which is adequate to the language, we assume that elementary propositions are described by the Hilbert space formalism consistently as it is explained in 1.3.1. We call this specific example of an underlying semantics for elementary propositions a *Hilbert space semantics.*"

We observe that for the application of the formal language a set of "properties" and operational methods to "prove or disprove" elementary propositions associated with these properties are essential. The physical meaning of a "property" in this context is given by the operational methods to "prove or disprove" the associated elementary proposition. The application depends on "an established theory" by means of which "a precise account of what a physical operation means can be given".

Since ABQM allows a phenomenological description of all experiments which can be used to define operational methods it must be possible to exhibit a connection between ABQM and the propositional language approach. In the sequel we shall propose two different operational methods in the framework ABQM. However, the details of these proposals are not yet worked out. Assuming the mathematical scheme of [7] § 5 we only suppose axiom AV 1.2s (but do not require in the moment axioms AV 1.1., AV 3, etc.): As the set of properties we take the set of exposed faces of K.

First proposal: For an exposed face C of K let $\hat{F}_1(C) = \{f : f \in \hat{F}$ and $K_1(\psi(f)) \subset C\}$ and $\hat{R}_1(C) = \{b \in \hat{R}$: there exists b_o with $(b_o, b) \in \hat{F}_1(C)\}$. As an experimental test for the elementary proposition "the system x has the property C" we propose the experimental statement

$$x \in M_{1r}(C) \underset{\text{def}}{=} \bigcup_{b \in \hat{R}_1(C)} b$$

Immediately, the following questions arise. What about the application of the formal language if the set of exposed faces of K is not an orthomodular lattice? Is axiom AV 1.1 together with AVid a sufficient condition that the formal language may be applied? Similar questions arrise for the second proposal.

Second proposal: For an exposed face $C \subset K$ let $\hat{Q}_1(C) = \{a : a \in \hat{Q}' \text{ and } \varphi(a) \in C\}$. As an experimental test for the elementary proposition "the system x has the property C" we propose the experimental statement

$$x \in M_1 \underset{\text{def}}{=} [\bigcup_{a \in \hat{Q}_1(C)} a] \cup [\bigcup_{b \in \hat{R}_1(C)} b].$$

We notice that the "physical problem" of the propositional language approach is removed to the problem of applicability of the language. If a formal language indeed has a region of applicability it might perhaps prove necessary to invent further languages when new regions of reality are discovered as e.g. the region of this so-called elementary particles.

We don't know whether this view is controversial but there seems to be a difference in the basic conception of a physical theory on which ABQM is founded [8] and the propositional language approach. This difference appears in the following sentences already quoted above:

"The operational proof procedures for elementary propositions which belong to the underlying semantics of the language to be established must be described precisely and completely by the language in the final stage when the language is established. This demand with respect to an operational approach to a physical theory is called semantical consistency or self-consistency".

From the point of view of ABQM and the underlying conception of a physical theory [8] such a problem does not exist since every theory is ultimately based on a very simple language by which we express that a fact is established. Questions corresponding to the so-called semantical consistency only appear as "intertheoretical problems" as described in [8] §§ 8 and 10.5.

Summarizing the discussion of this section LQP is not in contradiction to the other approaches and the connection between LQP and the other approaches can be found in the applicability of the formal language. Differences seem to appear with respect to in the philosophical background of the considered physical theory.

References

[1] C. Piron: Foundations of Quantum Physics. Benjamin 1976.
[2] C. Piron: La description d'un système physique et le présupposé de la théorie classique; Annales de la fondation Louis de Broglie, Vol. 3 n. 3, 1978.
[3] C. H. Randall, D. J. Foulis: Empirical Logic And Tensor Products, this volume.
[4] C. H. Randall, D. J. Foulis: Operational Statistics And Tensor Products, this volume.
[5] P. Mittelstaedt: The Concepts of Truth, Possibility And Probability in the Language of Quantum Physics, this volume.
[6] E.W. Stachow: The Propositional Language of Quantum Physics, this volume.
[7] G. Ludwig: An Axiomatic Basis of Quantum Mechanics, this volume.
[8] G. Ludwig: Grundstrukturen einer physikalischen Theorie, Springer 1979.
[9] C. Piron: Ideal Measurement And Probability in Quantum Mechanics, Symposion on Logic, Measurement And Probability in Quantum Mechanics, June 1979 (to be published).
[10] G. Ludwig: Foundations of Quantum Mechanics, Springer (to appear).
[11] E.W. Stachow: On the Logical Foundation of Quantum Mechanics (Habilitationsschrift, Univ. Köln 1980).

REFERENCES TO CONTRIBUTIONS OF THE COLLOQUIUM NOT INCLUDED IN THIS VOLUME

C. A. Hein (M.I.T. Lincoln Lab., Lexington):
Information Theory and Quantum Theory in Operational Statistics,
a summary of
C. A. Hein: Entropy in Operational Statistics and Quantum Logic, Found. Phys. 9, 751 (1979)

F. Jenč (Fachbereich Physik, Universität Marburg):
The CA (conceptual analysis) method in theories of microchannels and its application for the determination of the fundamental structure of quantum theory,
a summary of
F. Jenč: The CA (conceptual analysis) Method in Theories of Microchannels, Found. Phys. 9, 589, 707 and 897 (1979)

W. Ochs (Institut für theoretische Physik, Universität München):
No-Go-Theorems for Hidden-Variable Theories, to be published.